Fernanda Liégina da S. Fernandes
Beatriz S. Ovruski de Ceballos
Ivanise Gomes

Water, Health and Environmental Sustainability in the Brazilian Semi-arid Region

Fernanda Liégina da S. Fernandes
Beatriz S. Ovruski de Ceballos
Ivanise Gomes

Water, Health and Environmental Sustainability in the Brazilian Semi-arid Region

The case of Chã da Pia

ScienciaScripts

Imprint
Any brand names and product names mentioned in this book are subject to trademark, brand or patent protection and are trademarks or registered trademarks of their respective holders. The use of brand names, product names, common names, trade names, product descriptions etc. even without a particular marking in this work is in no way to be construed to mean that such names may be regarded as unrestricted in respect of trademark and brand protection legislation and could thus be used by anyone.

Cover image: www.ingimage.com

This book is a translation from the original published under ISBN 978-3-330-76476-7.

Publisher:
Sciencia Scripts
is a trademark of
Dodo Books Indian Ocean Ltd. and OmniScriptum S.R.L publishing group

120 High Road, East Finchley, London, N2 9ED, United Kingdom
Str. Armeneasca 28/1, office 1, Chisinau MD-2012, Republic of Moldova, Europe
Managing Directors: Ieva Konstantinova, Victoria Ursu
info@omniscriptum.com

Printed at: see last page
ISBN: 978-620-8-50375-8

Dedication

"I thank thee, because thou hast fashioned me by a marvellous way: thy works are marvellous, and my soul knoweth it right well" SI 139:14.

Thank you

To our families for the love and joy that accompany and encourage us in our daily lives.

To the artisans of Chã da Pia, who opened their doors to receive our team and kindly co-operated with the research, and for the teachings received that can't be found in any book.

To all the colleagues who contributed directly or indirectly to the research, especially Tércio José, a dear work companion who was present at all the campaigns and whose help was essential in the field.

The State University of Paraíba (UEPB) and the National Council for Scientific and Technological Development (CNPq), for the grants awarded under the Scientific Initiation Programme (PIBIC).

SUMMARY

This work was carried out in the Chã da Pia community (Areia/PB) during a period of long drought. The families interviewed are farmers who also produce utilitarian clay pieces for their own consumption and for sale. The aim of the work was to study the origin and management of the multiple waters destined for multiple uses in this community, to analyse the users' perception of the quality of these waters and their association with health risks and the impacts of the exploitative uses of water, soil and vegetation in the collection of materials for the manufacture of clay pieces. Ten families of farmers/artisans were interviewed (25% of the families in the community). In times of normal annual rainfall, agriculture is practised during this period and in the dry season the clay pieces are made, which generates extra income for the families. The water used is distributed by water tankers, with varying origins and qualities: 1) drinking water distributed by the water tanker operation; 2) treated water or water from untreated reservoirs supplied by local councils and 3) water of unknown origin distributed by private "pipeiros". This water is stored in cisterns, most of which are part of the PI MC programme - One Million Rural Cisterns, started by ASA in 2002. Some cisterns contain mixed water from different sources and are used for human consumption and household hygiene, which may indicate a lack of awareness of the importance of water quality and health risks (transmission of water-borne diseases). The extraction of clay "the clay of crockery" and the cutting down of plants to produce firewood for the firing of clay pieces have a strong environmental impact (soil and vegetation). The interviewees denied that it had any environmental impact and some justified this by saying: "it's hard soil that's no good", "it's scrubland, it grows on its own", or they made no comment at all. They had doubts or did not consider the simultaneous management of water of different qualities to be a health risk. The multiple waters could be of even more diverse origins and better utilised with alternative technologies that do not exist in this community, such as cisterns and underground dams, facilitating multiple uses during droughts.

CONTENTS

CHAPTER 1

INTRODUCTION

Water has always been and continues to be an indispensable resource for society. It is present in our bodies (70 per cent of the human body is made up of water), in the food we eat - 70 per cent of the water used is in agricultural production, in our daily activities (bathing, washing our hands, washing dishes and clothes), in our leisure activities (bathing in a reservoir or swimming pool), in the production of electricity.

In the last 60 to 70 years, population growth has been increasing at a rapid pace and consequently the amount of water needed for multiple uses (public supply, irrigation, energy production, navigation, recreation, tourism, fishing, mining, among others); but water uses have also become more diversified and more intense, which has resulted in impacts such as environmental degradation and pollution, making the availability of good water scarce in many regions (TUNDISI, 2005).

Since the middle of the 20th century, the world has been going through a water crisis that threatens humanity, increases water-borne diseases, produces economic and social stress and increases inequalities between regions and countries (TUNDISI, 2005).

In Brazil, the availability of fresh water is between 12% and 16% and has a heterogeneous distribution, generating various inequalities between the country's regions, such as the North, which concentrates almost 70% of this resource for a minority of the population (6.8%), while in the Northeast the availability of fresh water is only 3.3% of the total available for a population corresponding to 28.9% of the total inhabitants of the Brazilian territory.

The inhabitants of Brazil's semi-arid region, especially rural communities, are the most affected by the scarcity of good quality water and must resort to different sources to meet their needs, such as dams, wells and reservoirs, which, when they do have water, is of dubious quality for human consumption and other noble uses.

Brazil's northeast is currently experiencing one of the worst droughts in 30 years, and most reservoirs have already collapsed. The most affected states are Pernambuco, Paraíba, Ceará and Rio Grande do Norte (BRASIL, 2016). According to the United Nations, by 2030 a large part of the world's population will live in regions with water scarcity. In the northeastern semi-arid region, the lack of water is a fact of life for the population, especially in rural communities which, due to the prolonged drought of more than five years, have had their daily activities affected, as well as agricultural and livestock activities.

It's not just today that the Northeast is facing serious water scarcity problems: historically, the region has been characterised by long, cyclically repeated droughts that have devastated large areas and numerous communities, causing their populations to flee due to thirst, hunger and poverty. The lack of water is linked to a number of factors that have a negative impact on socio-economic growth and development. Among them is the irregularity of the rainfall; although the average rainfall is 800 mm per year, it is concentrated in just a few months of the year, and together with the high temperatures and evaporation rates (which are up to three times higher than the annual rainfall), this creates conditions of semi-aridity and uncertainty about the arrival of the next waters. However, these difficulties can be mitigated and even overcome with specific public policies for the region that include alternative technologies for collecting and preserving water during the rains for use in times of scarcity.

Historically, the semi-arid region has had major social inequalities associated with droughts and the consequent difficulties in socio-economic development: more than half (59.1%) of Brazilians living in extreme poverty reside in the semi-arid Northeast. Of these, 52.5 per cent are in rural areas (IBGE, 2010). The Human Development Index in the region ranges from low to very low (0.66 to 0.59) compared to the Brazilian average (0.727). Around half of the Semi-Arid population has no income or only government benefits as their sole source of income (IBGE, 2010).

Because of these striking characteristics, a false idea has been formed about the semi-arid region, which is seen as a permanently arduous place, with a difficult life, punished by drought, with images of dead animals, cracked earth, pregnant women and children searching for water on long walks. This image has been changed over the last three decades or so by a new conception, a new model that has been emerging in the semi-arid region: a re-signification that has been taking place and generating a new outlook: from an unviable space to a viable space, a new paradigm of connection between man and the environment in which he lives has emerged (SANTOS, 2012).

The issue of "coexistence with the semi-arid region" has been discussed by a number of authors who highlight the importance of this new paradigm and appears at a time when the region's inhabitants realise that public policies are not capable of solving many of the semi-arid region's problems. The paradigm shift in the way we look at the region shows that it is feasible to (con)live and produce, respecting local knowledge and culture, seeking new techniques and appropriate methodologies that transform the scenario and value ecological potential (SANTOS, 2012). Given this new concept, measures involving access to water, alternatives based on technologies and strategies for living with the SAB have been prioritised in sustainable development programmes in the region (SILVA, 2003).

A new identity is acquired by the population of SAB. New popular projects developed by civil society began to emerge from experiences with and in the region. These are Social Technologies (ST), based on spreading solutions to regional problems, considered low-cost and distributed in a wide variety of areas such as education, the environment, health and agriculture (JÚNIOR; PEDREIRA, 2004).

The One Million Rural Cisterns Programme (P1MC) and One Land and Two Waters (Pl+2) are both technological (and social) solutions that seek to guarantee rural families in the semi-arid region the sustainability of their daily and socio-economic activities even during drought events. They are programmes developed by ASA - Articulação no Semiárido, which defends, promotes and puts into practice, through public policies, coexistence with the semi-arid region (ASA, 2016).

PI MC aims to build one million plate cisterns and promote the decentralisation of water supply structures. The pre-moulded plate cistern adopted by PI MC was devised by a bricklayer from Sergipe known as Seu Nel. It was adopted by PI MC due to its ease of construction, long lifespan and low cost (R$1,800.00 to 2,300.00). With a volume of 16,000 litres, it supplies each member of a family of five with 8 to 9 litres of good water a day for 12 months (ASA, 2016). Storing rainwater in cisterns in the backyard has allowed easy access to good quality water for consumption, as well as democratising water, since the family itself manages it. Improvements in quality of life have made it possible to reduce water-related illnesses, increase school attendance among children and adults, reduce women's workload, pay more attention to their children and, for women, invest in their own education and community activities.

As a social technology, PI MC goes beyond the act of supplying water; it is a stimulus to community organisation and the conquest of citizenship, including the defence of the rights to access good quality water in adequate quantity for a dignified life, education, food, housing and citizenship (ASA, 2016). With the reduction in water-borne diseases, infant mortality has decreased, as have the long walks to collect water. Greater independence from local politicians and landlords is established, politicians and colonels who have always used the basic needs of the inhabitants in their favour, especially that of having water on their land (MALVEZZI, 2007).

CHAPTER 2

OBJECTIVES

2.1. General objective

The aim was to study, in a community of farmers and potters ("loiceiras") in the semi-arid region of Paraíba, the origins, forms of storage and multiple uses of water from various sources used in numerous activities, to learn about water storage technologies, which are fundamental for food and water security in rural areas, as well as to analyse the perception of users about the possible impacts of exploitative uses of the ecosystem (water, soil and vegetation) that can affect the sustainability of the environment where they live and earn their living and constitute risks to the community's health.

2.2. Specific objectives

a) In a community in the semi-arid region of Paraíba, learn about water sources and "waterways" and how they are managed, from those used for drinking to those "used up" in the home, in family farming and in the production of clay or ceramic pieces.

b) Analysing users' perception of the impacts that can be caused by exploitative use of the ecosystem (water, vegetation, soil) where they live and carry out their activities.

c) To verify possible health risks related to the handling of this water and the users' perception of this possibility.

CHAPTER 3

THEORETICAL FRAMEWORK

3.1. Water, present and future.

Water is a strategic resource, it is an essential commodity for life and is used in a variety of ways to meet all human needs, whether for consumption, hygiene, food production, recreation, energy generation, among many others. In the human body, water accounts for around 70 per cent of the body's mass and is the vehicle and solvent that allows the exchange of all the substances necessary for life, taking part in maintaining body temperature, among other functions (BRASIL, 2015).

In the words of Pielou (1998) *apud* Tundisi and Matsumura-Tundisi (2011), "Water is much more than a natural resource. It is an integral part of our planet. It has been present for billions of years, and is part of the functional dynamics of nature."

Planet Earth, known to many as planet water, is 75% water and the remaining 25% land (VIEIRA, 2006). It is estimated that the planet contains 1.37 billion km^3 of water (1.37x10 m) that recirculates through the hydrological cycle. Of this volume, 97.5% is concentrated in the oceans and the remaining 2.5% is fresh water. Of this, more than 68.9% is concentrated in the polar ice caps and glaciers, but it is not yet possible to collect this water. In addition, 29.9 per cent is groundwater, 0.3 per cent in rivers and lakes and 0.9 per cent in other types of reservoirs (REBOUÇAS, 2001).

As a renewable resource through the natural water cycle, springs were considered inexhaustible and over the centuries water management was neglected in both quantitative and qualitative aspects (MALVEZZI, 2007). However, at the beginning of human settlements and soon afterwards, with the construction of the first cities, which were always close to rivers in order to have water sources and to discharge and dilute waste, the impacts were reduced due to the smaller volume of sewage discharged compared to the dilution capacity of the receiving bodies (PHILIPPI; MARTINS, 2005). Quality was assessed according to its organoleptic characteristics or aesthetic aspects, such as visual appearance, taste and odour, and was generally of good quality (CEBALLOS, DANIEL, BASTOS. 2009).

The rapid population growth of recent decades has altered the availability of fresh water, which is becoming scarce due to increased consumption and the consequent greater demand to satisfy the population's needs for food, industrialised products and other inputs. In 1900, the volume of water consumed in the world was around $580x10^9$ m^3; in 1950 it rose to $1,400x10^9$ m^3, i.e. it increased 2.4 times in 50 years; in 2000 it was $4,000x10^9$ m^3, i.e. in the following 50 years consumption increased 2.8 times. According to the UN, it is likely that in 2025 it will rise to $5,200x10^9$ m^3, which means a 2.3-fold increase in 25 years. Of all activities, irrigated agriculture consumes the most water in the world, around 70 per cent, with 22 per cent used in industry and 8 per cent for domestic use (MMA, 2016).

The United Nations Children's Fund (UNICEF) and the World Health Organisation (WHO) reveal that around 40% of the world's population (2.6 billion people) do not have basic sanitation services and point out that one in six people (around 1.1 billion people in 31 countries) do not have an adequate water supply system. United Nations projections indicate that, if the trend continues, by 2050 more than 45% of the world's population will be living in countries that cannot guarantee the minimum daily quota of 40 to 50 litres.person.day^{-1}. On the other hand, 91 per cent of the world's population has seen improvements in the quality of their drinking water. In terms of basic sanitation, one third of the world's population does not have access to these services, including drinking water, according to UNICEF and the World Health Organisation (ONUBR, 2015).

The WHO considers that decent hygiene requires at least 20 litres of good quality water per person.day^{-1}. The UN indicates that around 40 litres per person per day^{-1} are needed for drinking, personal hygiene, washing clothes, cooking, etc. In practice, there is a great disparity in the availability of water per person per day around the world and in regional differences within each country. It can be seen that water consumption is related to the country's level of economic development and the income level of its inhabitants: the United States has an average daily consumption of 570 litres.person.day^{-1}, in Europe the average is 350 to 200 litres.person.day$^{(-1)}$)', in Brazil it is around 185 litres.person.day$^{(-1)}$), in India it varies between 135 and 15 litres.person.day$^{(-1)}$) and in Haiti from 35 to 15 litres.person.day$^{(-1)}$. Developing or underdeveloped countries always have lower consumption.

Another component that limits the availability of fresh water is its quality. Over the centuries, the complexity of man's multiple uses of water has greatly increased, with its effects being the pollution and degradation of water quality in most water bodies, including water from springs intended for human consumption after potabilisation treatment (TUNDISI;

MATSUMURA-TUNDISI, 2011). Among the anthropogenic impacts on the drainage basin, discharges of untreated or poorly treated domestic and industrial sewage are the main cause of altered surface water quality and the cause of its eutrophication, which is expressed in blooms of cyanobacteria, algae and macrophytes. The water source loses its landscape and recreational value, and water treatment becomes more complex and costly; in addition, cyanobacteria and algae produce substances that cause an unpleasant taste and odour, and the former synthesise and release toxins that affect the liver, nervous system and skin; all of which are difficult to remove from water using conventional treatment methods (Dl BERNARDO; DANTAS, 2005).

With an area of approximately 8,514,876 km2 and a population of more than 204 million inhabitants based on data up to 1 July 2015 (IBGE 2015), Brazil stands out in the global context for its water capacity: it has between 13% and 16% of the fresh water available on the planet. However, the availability of these resources is not uniform. Around 70 per cent of the country's fresh water is concentrated in the Amazon basin, which accounts for only 6.87 per cent of the country's pollution; in the southeast, with more than 43 per cent of Brazil's inhabitants, the available water resources account for only 6 per cent of this total and in the semi-arid northeast, with a population of 28.9 per cent, the available fresh water corresponds to only 3.3 per cent of the total that exists in Brazilian territory. Data on population density and the availability of water resources are presented below.

Region	Population density (inhab/km^2)	Concentration of the country's water resources
North	4.12	68.5%
North-East	34,15	3,3%
Centre-West	8,75	15,7%
South East	86,92	6%
South	48,58	6,5%
	Source: IBGE / National Water Agency (2010)	

Table 1 - Distribution of water resources and demographic density in Brazil **Source:** IBGE/National Water Agency (2010).

Rebouças (1997) reports in Dossiê Nordeste that the hallmark of the 21st century would be "the water crisis", which would affect both the global and national scenarios. The distribution in table 01 highlights the national disparity. Brazil faces a number of challenges in guaranteeing an adequate water supply for the different regions. These challenges must be met with strategies that guarantee larger volumes of water and less waste, integrating new sources such as: reuse, the sustainable exploitation of underground aquifers, soil protection and recovery with actions aimed at promoting the recovery of vegetation and riparian vegetation removed from the site, improving the absorption of water by the soil and reducing siltation, improving the quality of water in the springs (TUNDISI2005).

3.2. Water in the Brazilian semi-arid region.

The Brazilian semi-arid region covers an area of 982,566 square kilometres, extending over nine states and involving 1,135 municipalities from Piauí to the north of Minas Gerais. The region is home to more than 28.9% of Brazil's inhabitants, around 40% of whom live in rural areas, and has only 3.3% of the country's water resources (ASA, 2016).

A large part of this 40 per cent corresponds to the inhabitants of scattered rural communities without access to drinking water, who are supplied with rainwater (when it rains), stored in pots and cisterns and by water tankers, a form of water distribution that is integrated into the semi-arid landscape and which intensifies during inter-annual droughts and especially during prolonged droughts of more than a year's duration. Water tanker suppliers don't always distribute water of a quality that is safe for human consumption. Water from dams, cisterns and reservoirs is also used, generally for "household use", washing floors, toilets and clothes.

The landscape that predominates in the region is the caatinga biome or "white forest" in the indigenous language, which is exclusively Brazilian, with a wide variety of endemic species, its trees and shrubs are medium to small in size, with an abundance of xerophytic and deciduous plants, such as Mandacaru, Xiquexique, Juazeiro and Aroeira (PORTO et al. 2011). It stands out for its ability to survive the dry seasons. When the first rains fall, the caatinga turns an exuberant green and loses the grey colour it acquired during the dry months (figure 2)

Figura 1. a) Mandacaru; **b)** Aroeira; **c)** Xiquexique; d) Juazeiro **Source:** INSA, 2015; APNE/CNIP

Figure 2 - Farm in Piauí before and after the rainy season in the region.
Source: cidadeverde. com/noticias/212017/chuvas-mudam-paisagem-no-piaui-confira- os-registros, (2016).

The municipalities that make up the semi-arid region of north-eastern Brazil, according to the Ministry of National Integration (BRASIL, 2005), which drew up the new demarcation of the semi-arid region (figure 3), fulfil the following criteria:

1. Average annual rainfall of less than 800 millimetres;

2. aridity index of up to 0.5 calculated by the water balance that relates rainfall and potential evapotranspiration in the period between 1961 and 1990;

3. Risk of drought greater than 60 per cent, based on the period between 1970 and 1990.

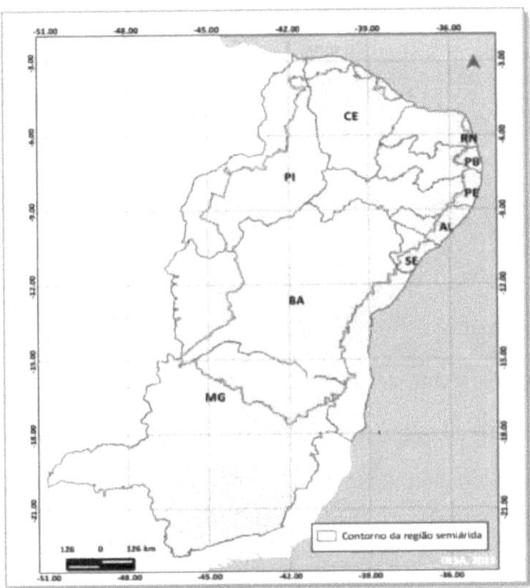

Figure 3. The Brazilian semi-arid region.
Source: INSA. (2014)

The Northeast has two climatic seasons: one very dry and the other moderately rainy. With an average of 800 mm of rainfall per year, there should be no shortage of water in the SAB, but the rains are concentrated in a few months of the year (four to six months), which the region's inhabitants refer to as winter. It is not uncommon for this "winter" to be interrupted by veranicos without rain for weeks or even months. This climate variability, the uncertainty of rainfall and the lack of proper management of the available water are the main causes of the "lack of water in the region". As a result, the northeastern semi-arid region faces problems related to the "lack of water". These include the limitation of agricultural activities, extensive livestock farming, difficulties in sustaining large-scale industrial activities, among others.

According to Marengo (2011), one factor that contributes to rainfall variability in the region is the EI Nino phenomenon, which alters the normal patterns of sea surface temperature and trade winds in the equatorial Pacific region. These phenomena modify the regional climate and the global climate, with changes in winds worldwide and in rainfall regimes in tropical and mid-latitude regions. According to the same author, they can also cause drought in the north-east of Brazil by interfering with rainfall, known as a dipole or Tropical Atlantic gradient, which is characterised by variations in the temperature of the Atlantic Ocean and causes cold fronts coming from the south of the country to be blocked.

Another factor that contributes to the semi-aridity of the region is the geological formations: made up of two structures: the crystalline bedrock that makes up around 60% of the semi-arid region and the sedimentary rocks that affect the availability of water, especially groundwater (ALBUQUERQUE; REGO 2011). The crystalline soils in much of the region are practically impermeable and limit the supply of aquifers so that around 90 per cent of rainwater is not used due to rapid surface run-off and high evaporation (ASA, 2015). Even though they were built to meet the need for water, many wells have low flow rates and quality that limits their use, in particular the concentration of salts exceeds the recommended standard for human consumption of 250 mg.L^{-1} and is a consequence of the influences of geological formations (BRASIL, 2011; CIRILO, 2011).

The variability of rainfall is the cause of the predominance of intermittent rivers, and the main sources of water in the region are the surface reservoirs built in the headwaters of rivers, which vary from low to high capacity (ARAÚJO, 2012; ALBUQUERQUE, RÊGO, 2011).

Built to balance water availability and satisfy the demand for water in the region, the "weirs" have a long hydraulic detention time, a wide water mirror and are generally not very deep, and are exposed to evaporation, which exceeds rainfall for several months. In years with heavy rainfall, the water that overflows (when the "weir bleeds," in the popular parlance) is lost to the sea (CIRILO, 2011). For 8 years prior to the current drought, which began in the last months of 2012, all of Paraíba's reservoirs bled with water levels of more than 2 metres (AESA, 2016) - data from 2004 to 2012 - without adequate management being sought to make use of this water. In turn, hundreds of low-capacity dams and clay ponds are built without planning, sometimes upstream of the large dams that are supposed to supply human settlements, and end up salinising the flow of water that is supposed to fill the large dam downstream.

Over the last 14 years, a large-scale and successful social technology programme has been developed in the north-east of Brazil to supply communities scattered across the semi-arid region with good water. The technique is based on recovering and reinventing rainwater storage in household tanks called cisterns.

This is the first programme developed by ASA, which began in 2002 and is called the

One Million Rural Cisterns Programme - "P1MC". The programme aims to provide good quality drinking water to the rural population without access to drinking water and to strengthen the decentralisation and democratisation of water by storing rainwater in cisterns of sufficient volume for use for a year by families with up to five members (ASA, 2016). By using the rainwater stored in the cisterns, these families are no longer dependent on the water from dams or reservoirs, sometimes built on the land of large farmers, but can manage their own water. As of 18 April 2017, 597,403 cisterns have already been built in the semi-arid northeast; considering an average of five inhabitants per household, more than 2.5 million people have benefited from good water (ASA, 2017).

Collecting and storing rainwater is an ancient technique that has been used for centuries in regions where rainfall is scarce or irregular throughout the year (XAVIER, et al, 2012). Various peoples around the world use popular techniques to capture and manage rainwater in different regions. Some of these techniques have been and continue to be used by people in arid and semi-arid regions, such as Iran, where it is possible to find stone and lime mortar tanks with a tower for the passage of air that cools *the* water, known as *'cibcinbcn"b* (figure 4); the Maya who lived on the Yucatan peninsula, near the town of Oxkutzcab, at the foot of Mount Puuc, had cisterns that could store from 20.These cisterns were dug into the limestone subsoil and lined with waterproof plaster (GNADLINGER, 2011).

Figura 4. Abanbars, Iran's traditional community water catchment system. **Source:** Gnadlinger, (2004)

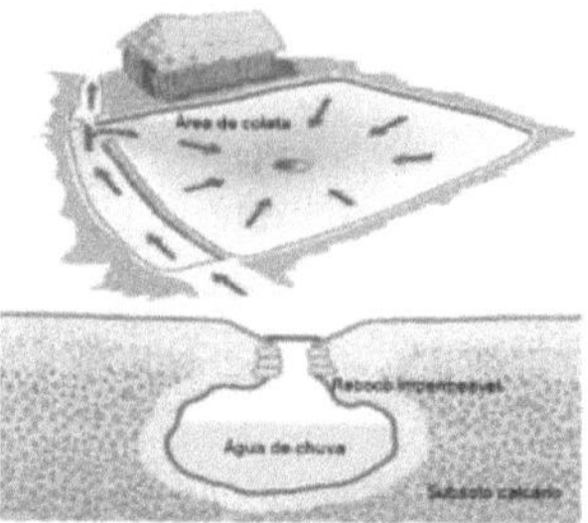

Figure 5. Chultuns: underground reservoirs of the Maya
Source: Gnadlinger, (2011)

Two thousand years ago, on the Loess Plateau in China's Gansu Province, there was a system of cisterns and tanks that collected rainwater. In the 11th century, in Southeast Asia, the Krner people developed a water supply system to irrigate large areas of rice through a system of reservoirs, canals and dams. In Thailand, a drinking water collection programme was developed in the 1980s and 1990s using cement slab jugs (figure 6) that can hold up to 2,000 litres (GNADLINGER, 2011).

Figure 6. Jugs with rainwater in Thailand
Source: Gnadlinger, 2009.

3.3. Main public policies to combat drought.

For decades, the Brazilian Northeast has faced several droughts without any assistance from public policies to alleviate or even solve the problem, so its history is intertwined with the drought events that have devastated the region and caused great damage. The first major drought recorded in Brazil was in 1583, when the Jesuit Femão Cardim arrived in the country and reported the drought that devastated the state of Bahia, forcing part of the population to move to the coast and reducing sugar production in the mills (POMPONET, 2009).

In 1729 there were the first official records, when many slaves died of starvation and the mills came to a standstill. There were several other droughts during the empire, but the one in 1877 stood out and became known as the "great drought", being highlighted throughout Brazil because it affected half a million northeasterners with hunger, death, disease and exodus. As a result, debates began to discuss drought as a problem and suggestions were made as to what actions would be necessary to alleviate its effects (CAMPOS, 2014; LUCCHESI, 2016).

The governmental actions chosen to resolve the issue of droughts were established according to three characteristics, according to Silva (2003): 1) of an emergency nature, with fragmented programmes developed at times of public disaster; 2) of emergency actions that sustained the "drought industry"; and, 3) of hydraulic solutions, with water works, favouring contractors and large rural property.

Ferreira (2008) considers that the "drought industry" can be analysed in two ways: 1) locally - in which the embezzlement of funds was carried out by members of the victim relief committees, and 2) nationally - in which Northeastern representatives benefited from using droughts as an excuse to get government investment in the region.

The same author emphasises that public aid became resources for electoral arrangements, so that cattle breeders and dam owners remained responsible for distributing (and diverting) the economic aid that came to the needy population; the funds earmarked for the construction of "anti-drought" works were invested without any planning or care for their completion (FERREIRA, 2008).

The situation was no different in Paraíba, where the state remained under oligarchic power, just like the whole of the Northeast. Almost two centuries after the "great drought", in October 1909, the government of Nilo Peçanha created the first national institution to combat drought, the Inspectorate of Works Against Droughts (IOCS), the first body to study the problems of the semi-arid region (DNOCS, 2016). A few years later, in 1919, it became IFOCS (Federal Inspectorate) under the presidency of Epitácio Pessoa, who began to pay more attention to the drought, and in 1945 under the government of Getúlio Vargas it was renamed DNOCS - National Department of Works Against Droughts (POMPONET, 2009; LUCCHESI, 2016).

With the creation of the IOCS, the owners of power in the northeast lost control of the distribution of aid to the victims of the drought and also of the privileges they had acquired. But the privileges did not come to an end, they just took a new direction and the owners of cattle and mills in the region continued to benefit. The new methods used are discussed by Ferreira (2008), who highlights: the construction of roads close to their properties; public dams on private properties that had never been expropriated; the use of sheds to sell food, which guaranteed the permanence of the labour force and prevented it from becoming more expensive in the period after the drought. The fight against drought became a big business, sustaining the power relations belonging to the social class of landowners who benefited from the works carried out in the semi-arid region.

From the beginning of the IOCS until the creation of SUDENE, construction work in the region concentrated on dams, roads, bridges, harbours, railways, hospitals, airfields, electricity grids and hydroelectric power stations (DNOCS, 2016).

In Paraíba, the government of Epitácio Pessoa invested heavily in public reservoirs and road construction. However, after the end of his term, various works were paralysed and only restarted under José Américo in the 1950s (LUCCHESI, 2016).

In 1958, another drought hit the Northeast, killing half a million people with hunger. The government of Juscelino Kubitschek had been pursuing a developmentalist policy in the country and placed in the hands of Celso Furtado, a Paraíba native from Pombal, the mission of creating a plan for the Northeast, which at that time was lagging far behind the South and Southeast regions (TAVARES et al.; 1998). Celso Furtado then visited the entire Northeast and organised a small team to draw up the project to create SUDENE. In 1959, under the direction of Celso Furtado, the Superintendence for the Development of the Northeast (SUDENE) was created to promote and coordinate the development of the region (TAVARES et al.; 1998).

With the creation of SUDENE, DNOCS would be controlled by the agency, which did not please the majority of the northeastern elites, who were against the proposal, because with control of DNOCS, the diversion of resources that supported the "drought industry" would cease to exist. In 1964, under the military government, SUDENE lost its autonomy and had its activities reduced, and its director was removed with his political rights cancelled (ANDRADE; NUNES 2014).

In May 2001, SUDENE was closed down following accusations that it was favouring a clientele. According to the analysis of sociologist Francisco de Oliveira, the agency failed to solve the region's structural problems, especially the persistence of models of poverty and the continued migration of its inhabitants (migrants from the droughts) (OLIVEIRA, 2015).

3.4. Living in the semi-arid region: sustainability and social technologies

At the turn of the 20th to the 21st century, the semi-arid region was characterised by the action and effect of public policies both to combat drought, with actions to build and renovate dams and reservoirs, and to live together, with programmes to combat rural poverty, family farming, crop grants, credits for farmers and systems for collecting and storing rainwater in cisterns (ANDRADE; NUNES, 2014).

A discourse is emerging about the reality of the region with sustainable development alternatives for the Brazilian semi-arid region (SILVA, 2003). "A change in the model that has been reflected in public policies", which encourages a new look at the qualities that the region offers, acceptance and appreciation of its physical and climatic characteristics and its potential in flora and fauna, among many others (SANTOS et al., 2013).

It was during the 1980s and 1990s that various non-governmental organisations (NGOs) and some public institutions "began to develop proposals and experiment with alternatives based on the idea that it is possible and necessary to live with and alongside drought and the semi-arid region" (SANTOS et al., 2013). According to the same authors, the proposed actions developed in the region by these groups have made it possible to unite state and society, resulting in coexistence with the place in which society participates in the construction and implementation of public policies. Customs and traditions can be maintained, strengthening local identity and increasing the connection with the land (COSTA; DIAS, 2013).

The Training and Mobilisation Programme for Coexistence with the Semi-Arid - One Million Cisterns Programme (PIMC), institutionalised in 2003, is the result of this process (SANTOS 2012). Created by Articulação do Semiárido - ASA in partnership with the Ministry of Social Development and Fight against Hunger and many other national and international partners, the P1MC

seeks to provide good quality water for human consumption by collecting rainwater in reservoirs or cisterns built by the community itself using the pre-moulded concrete slab method (figure 7) (ASA 2015, ASA, 2016).

The programme's cistern is round and built close to the beneficiary family's home. This type of cistern is made from pre-moulded cement slabs, covered with galvanised steel wire mesh and then plastered. Part of the reservoir is buried to give it firmness and withstand the pressure of the water to be accumulated. The roof is cone-shaped and is also made from cement slabs. Once the cistern is ready, it is given an external coat of white paint to protect the plaster and reduce the temperature of the water (GNADLINGER, 2011; ASA 2016).

Figure 7. Cement slab cistern used in the P1MC.
Source: ASA, (2016)

The pre-moulded slab cistern was the brainchild of a bricklayer from Sergipe, Seu Nel, and quickly spread throughout the semi-arid region. It was adopted by the P1MC due to its ease of construction, long life and low cost (R$1,800.00 to 2,300.00). Its volume of 16,000 litres supplies each member of a family of five with 8 to 9 litres of good water a day for 12 months. The improvements in quality of life with the adoption of Rainwater Catchment and Storage Systems in Cisterns (SCAAC) have made it possible to reduce water-borne diseases, especially among children under five, increase school attendance among children and adults, reduce women's workload, pay more attention to their children and invest in their own education.

As a social technology, PI MC goes beyond the act of supplying water; it is a stimulus for community organisation and the conquest of citizenship, including the defence of the rights of access to good quality water of adequate quantity for a dignified life, up-to-date education, proper nutrition, decent housing, among other achievements (ASA, 2015).

The Rainwater Catchment and Storage System in Cisterns (SCAAC) is made up of three elements: a) the catchment area (the roof), b) the ducts that receive the captured water (gutters) and the downpipes that lead the water to the third component, c) the reservoir (the cistern) (figure 8). This must have a manual pump to remove the water without contact with equipment that may be dirty and also without contact with the people removing the water, a lid in good condition to close the cistern and protect the water from dust and insects. There also needs to be a reservoir to divert the first rainwater that washes off the roofs and is the dirtiest - this is where the water is retained, preventing it from entering the cistern (HEIJNEN, 2015).

Figura 8. Rainwater Harvesting and Storage System used in the Brazilian semi-arid region.
Source: Lima, 2015

As a social technology, PI MC goes beyond the act of supplying water; it encourages community organisation and the achievement of citizenship, including the defence of the right to access good quality water in adequate quantity for a dignified life, reducing the suffering of the most fragile people, especially children and the elderly (ASA, 2015).

Another advantage of cisterns is that they can serve a large number of rural families without having a negative impact on the environment, when compared to large construction projects. However, there are some limitations that can make it difficult to achieve a greater reach: the PI MC adopts the same volume for all cisterns (16,000 L), which doesn't allow more local rainwater to be used, for example, from homes with larger roofs and a greater number of users (families with more than five members), so it is limited to the fixed volume of the cistern, which is calculated for catchment areas (roofs) of around 80 m^2. In fact, the average of five members is the most common in the region, and the catchment area adopted by PI MC is standard. Despite great efforts to meet basic water needs more easily, the SCAACs are unable to supply all families with the appropriate volumes, which according to the World Health Organisation should be at least 20 L.person.day'[1]. In order to achieve volumes of 9 to 13 L.person.day^{-1} for the calculated time of 8 to 12 months, the cisterns must be completely filled during the rainy season for use in the dry season or summer. Another criticism refers to the microbiological quality, which tends not to meet potability standards (Ordinance 2914/2011-MS), and must be carefully considered since the system needs care and maintenance as well as water management to preserve quality and even improve it, with disinfection at the final point of consumption that destroys contaminating microorganisms (ANDRADE NETO, 2013; ANDRADE, NUNES, 2014).

However, the criticisms do not take into account that PI MC's technical model can be adapted to capture a greater volume by increasing the roof area, if necessary, and by building larger capacity cisterns or more cisterns of the same volume using the same technology. The increase in costs versus advantages should take into account that more cisterns or larger cisterns would reduce expenses with emergency operations such as Pipa Operations, which are assistive, temporary, complex and expensive.

But family supply based solely on collecting and using rainwater is very fragile in the semi-arid region due to the variability of rainfall, prolonged droughts and unexpected summers, and can be even more fragile when it comes to preserving quality, since the care of the system and the preservation of quality is entirely the responsibility of the users, with economic limitations. Public policies that offer the inhabitants of the semi-arid region the conditions to build more than one cistern or a larger one would make it possible to accumulate more water free of charge, and self-manage these volumes by looking after the quality during the dry months and even during some years of drought. This proposal could be a golden and sustainable solution, efficient in terms of quantity and quality, but it needs the support of the Unified Health System, which should provide assistance by teaching hygienic care of water and also monitor quality and compliance with these measures. In this way, emergency operations such as Operation Pipa by the army and civil defence would be complementary to prevent water shortages in the cisterns throughout the drought and not to fill them up once they are empty.

In this context, social technologies (ST) have been created and developed in the communities themselves with the participation of local labour, using traditional knowledge, rescuing self-esteem and valuing local knowledge. A good example of ST is homemade serum: a little salt, sugar and boiled water that can prevent dehydration and death

(GUALDANI; et al., 2015).

ST was born from the concept of Appropriate Technology (AT), developed between 1924 and 1927 by Gandhi in India, who endeavoured to expand the manual spinning process with a spinning wheel as a way of fighting social injustices in India under the rule of the British Empire. In improving traditional techniques, Gandhi sought to adapt modern technology to Indian conditions and to stimulate scientific and technological research. With his work, he awakened the political consciousness of millions of Indians, exposing the need for self-determination and the transformation of Hindu society in a process of natural, endogenous growth and not as an external imposition (DAGNINO et al., 2004).

STs are basically simple, easy to use and based on the dissemination of sustainable solutions to problems related to food, education, energy, housing, income, water resources, health and the environment (FBB, 2015; MALVEZZI, 2007). According to Bava (2004), these are techniques and methodologies developed by and with the population, which have the power to transform and strengthen democracy and citizenship through the processes of building new paradigms and with new social actors, and are generators of improvements in quality of life.

CHAPTER 4

MATERIAL AND METHOD

4.1. Description of the study area

The work was carried out in the rural community of Chã da Pia, located in the north-western part of the municipality of Areia in the state of Paraíba in semi-arid north-eastern Brazil. This municipality, located between the Agreste Mesoregion and the Brejo Paraibano Microregion, has an area of 266,596 km and an estimated population of 23,288 inhabitants (IBGE, 2015). Its average Human Development Index (HDI) is 0.59 (IBGE, 2010).

The Chã da Pia community is well known in the Northeast and particularly in the state of Paraíba for its artisans who produce utilitarian ceramics: pots, stoves, jars and other household utensils. The community where the work was carried out is located between the agreste and the brejo, on the north-western edge, which transitions into the "Curimataú", the driest area of Agreste Paraibano. The geographical coordinates are: latitudes 6° 54' 15" and 6° 55' 6" South and longitudes 35° 46' 39" and 35° 47' 41" West of Greenwich.

The relief varies between undulating and gently undulating, with an altitude of between 484m and 552m. The climate is harsh, with average annual rainfall of 700 to 800mm, with minimums of 200mm and maximums of 1,000mm. The vegetation is xerophytic and typical of the caatinga. Figure 9 below shows the location of the community studied.

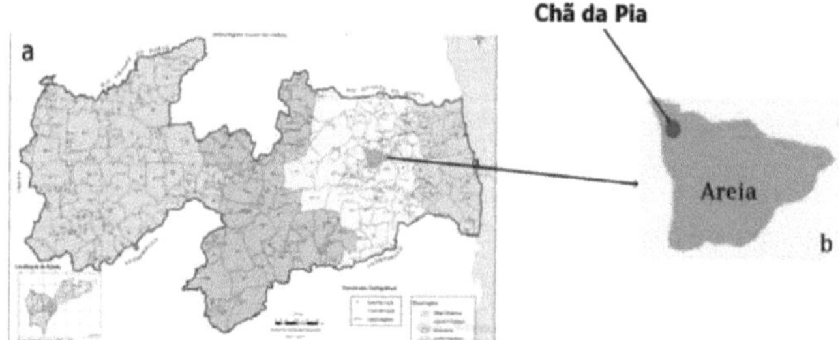

Figure 9 - Location of Chã da Pia on the state map: (a) State of Paraíba and mesoregions, location of the municipality of Areia; (b) Municipality of Areia, with Chã da Pia highlighted.
Source: Paraíba State Government, (2015).

Research period. The work lasted 12 months, from August 2014 to July 2015. Twelve field trips were made to carry out the different stages of the research from November 2014 to

July 2015.

Sample universe. It was made up of 10 families of potters who are also farmers, and who were selected from the group who agreed to take part in the research. The conditions for choosing these artisans were: the use of water from different sources for various uses, including family farming and the production of "earthenware". At the time of the research, the community had approximately 40 clay artisans, with a large predominance of women. The 10 artisans therefore represent 25% of the total.

4.2. Research stages

First stage:

The research began with a review of bibliographical material, a task that was carried out until the end of the work. At the same time, semi-structured questionnaires were prepared to be applied to families in the Chã da Pia community. These questionnaires have 10 central topics, totalling 114 questions, which covered topics that made it possible to get to know the habits of these artisans and their families, to carry out socio-economic characterisation: number of members, age, formal education, occupation, income, housing conditions, hygiene and basic sanitation - sewage disposal, origins, types and uses of water, water-borne diseases, SUS care through Community Health Agents (ACS) and Environmental Surveillance Agents (AVAs), and cultural aspects - characterisation of the work with "earthenware", from the removal, preparation, production, use and sale of the pieces, and family relationships around this activity..

Training was carried out in the laboratory of EXTRABES - Experimental Station for the Biological Treatment of Sanitary Sewage, belonging to the Paraíba Water and Sewage Company - CAGEPA, and loaned to the Paraíba State University - UEPB. This training aimed to familiarise them with the techniques used to analyse water quality parameters. Practicals were carried out using the techniques for quantifying bacteria that indicate faecal contamination (total *coliforms*, thermotolerant *coliforms and Escherichia coli)* using the filter membrane methodology and the Agar M-FC and chromogenic substrate - MUG culture media, turbidity using a turbidimeter, apparent and true colour using a spectrophotometer, among others.

Second Stage:

This stage involved field research, which was divided into three phases:

First phase - visits to get to know the community and the "loiceiros", who were later invited to take part in the research. During this phase, initial contacts were made to find out about the number of "loiceiros" working in the community, members per family and their activities. The site was geographically recognised, several families were visited and informal conversations were held with the artisans. The first questions, although informal, focused on the artisans' activities as potters and farmers. The aim was to accompany and understand the daily lives of the families as they lived in the region during the rainy and dry seasons; to learn about the multiple waters used for multiple uses, the stages of handicraft production of the clay pieces - from collecting the earth to prepare the clay to making and firing the "crockery" and selling it at the open markets in the nearest towns. Before starting the next phase, the information obtained in this first phase was analysed and systematised to help draw up the questionnaires that were applied to the "loiceiros" of Chã da Pia (nine "loceiras" and one "loicero").

The research participants signed the informed consent form *on the spot.* In this form, the researcher undertakes to safeguard the information obtained and the identity of the research participant, and also to respect the research participant's decision if they wish to leave the research at any time, in order to avoid any embarrassment. Annex 1 corresponds to this document.

Second phase - Called the **execution** phase**,** families who agreed to take part in the research were registered and after further informal conversations, open-ended interviews were carried out based on the semi-structured questionnaire. Information and opinions that emerged during the interviews that were not included in the questionnaire were recorded and greatly enriched the research. The interviews began in January 2015 and were completed in June of the same year, totalling 10 interviews (09 women and 01 male artisan). All the visits were unannounced, with the intention of meeting the families in their daily lives for a realistic *on-the-spot* observation and in search of spontaneity in the answers to the questions. The questionnaires were administered by a team of three people who had different tasks: one did the semi-structured interviews with the artisans and the other two were responsible for field observations. With the families' authorisation, visits were made to their homes (kitchens and bathrooms), backyards and neighbourhoods. The state of conservation of the catchment system and the cistern, the use of a bucket or pump to remove the water that is transported into the houses and its location in relation to the septic tank, the presence of loose animals near the reservoir, among other records, were observed. In the kitchen, the types of water stored and their packaging were checked.

Third phase. After the formal visit, follow-up visits were made at least twice to each family. These visits aimed to monitor any changes in the families and the types and uses of water. They were shorter visits, with a variety of questions on topics similar to those in the questionnaires applied beforehand and to detect any changes in the interval between visits, with an emphasis on possible changes in the origins of the water that supplied the families; cases of diarrhoea, changes in the production of clay pieces and in demands, conservation of the cistern, changes in the volume of water stored in it, family hygiene and the condition of the yard as the drought extended.

Fourth Stage. Database Organisation, Analysis and Interpretation. In the last stage, the database, which was fed throughout the research, was revised, completed and reorganised in Excel spreadsheets and the results analysed. The results were analysed comparatively, tables and graphs were drawn up and statistical analyses of measures of central tendency were carried out.

CHAPTER 5

RESULTS AND DISCUSSION

5.1. Multiple Waters and Multiple Uses: Management

The community of Chã da Pia does not have a mains water supply and needs various sources of water for all its needs. This water comes from rainfall, stored in cisterns, sinks (puddles dug naturally in the rocks) and dams. But it is the water tankers that dominate the semi-arid landscape during the drought and distribute the largest volumes of water. In fact, water tankers are present all year round, during the drought and even during the rainy season, although less frequently. Many rural communities in the semi-arid region depend on the water they distribute.

According to information from the National Household Sample Survey - PNAD/2014, in Brazil just over 33% of rural households have water supply networks with or without internal plumbing, and 67% use water from various sources, untreated or generally inadequately treated for human consumption. The table below shows this data.

Table 02 - Water supply per household in rural and urban areas in Brazil.

Area	Total number of households	Households connected to the grid			Other forms		
		With internal plumbing (%)	Without internal plumbing (%)	Total (%)	With internal plumbing (%)	Without internal plumbing (%)	Total (%)
Urban	57.641.000	93,37%	0,50%	93,87%	5,11%	1,02%	6,13%
Countryside	9.398 000	30,33%	3,09%	33,41%	46,57%	20,01%	66,59%
Total	67.039.000	84,53%	0,87%	85,40%	10,92%	3,68%	14,60%

Source: IBGE - PNAD 2014

The water available in the community during the research period, characterised by the long drought that began in 2012 and lasted until 2017, came from the water tanker operation controlled by the army and civil defence, which distributes treated (drinking) water in water tanker cars; water from municipal water tanker cars or those hired by local politicians, which generally don't transport treated water, but more often from reservoirs, and also from private water tanker cars, with water of unknown origin that users buy by paying between R$30.00 and R$8.000.00 for a volume of 6.000L to 8.000L.000L to 8,000L, between R$130.00 and R$170.00, depending on the quality and ease of access that the pipers find to collect and deliver this water. In the current drought, there are sparse rains and they are short-lived, and although there is little rainfall, it is possible to accumulate water in the cisterns, barrels and sinks. The latter have better quality water for use in preparing the clay for modelling the pieces produced by the "loiceiros", because they prevent the pieces from popping or breaking easily when they are fired.

These are the waters that must be managed by the community for multiple uses during the drought. In order to receive the water from the water trucks during the dry season, some of the PI MC cisterns were prepared for this purpose. Under the guidance of the army, they chose the ones that were in the best state of repair, and any cracks were covered over with cement, broken or cracked lids have been replaced or repaired, and they have been washed and painted with lime inside and out. They received good water from the pipa operation, of drinking water quality, which should have a chlorine residual of between 0.2 and 0.5 $mg.L^{-1}$. This water was reserved for drinking, cooking, personal hygiene and washing household utensils. Other cisterns, less well maintained, were reserved for storing water of dubious origins distributed by other water tankers that transport water from dams or other sources and are not treated. It is used to wash floors and some clothes, toilets and other facilities, to water animals, to irrigate flower pots, among other things. This distribution represents a categorisation of water according to its origin and quality, maintaining hygiene principles and the uses are defined according to quality. In other cisterns, water was mixed: leftover rainwater was mixed with water of lower quality, from dams for example. The three types of water were used for previously defined activities and the population recognised their uses.

During the rainy season, the community prioritises the use of water that is collected from the roofs and stored in cisterns, destined for noble uses (cooking, drinking) and that which has filled sources such as sinks, dams and reservoirs has been and is used for "spending" and for the production of clay pieces.

When preparing the clay for modelling the pieces ("crockery clay"), "waste" water is used, preferably water from sinks because it is not saline. Saline water makes it easier for the pieces to pop and makes them more fragile when cooking; an entire production can be lost. During the dry season, the craftswomen make use of good water if there is no accumulated water in the sinks.

The clay used to make crockery is called by the inhabitants of Chã da Pia "loiça clay" and the product "loiça de barro" or "loiça da pia" or simply "loiça" and the community does not consider them to be ceramics (ALVES, 2004). The greatest production of crockery takes place in the summer or dry season and less intensely during the rains or winter, as during this time the "loiceiros" prioritise agriculture, a common practice in the community. In this way, rainwater is managed for agriculture and to fill the cisterns and sinks. When it rains, the minimum is 200 mm per year or more and if the cisterns and catchment systems are in good condition, 16,000 litres are stored, enough to reduce water vulnerability in inter-annual droughts of eight months or even a year.

5.2 From "earthenware" to "earthenware"

Known as ceramics, earthenware and even "earthenware", the production of pots, bowls, jars and stoves is an activity that produces pieces that are made from clay, the oldest man-made material used for this purpose.

Pottery making is a very ancient art, created by the people of prehistoric times when they used it for domestic purposes and religious rituals. Many of the objects they modelled were used to preserve seeds and grains, to transport and store water, to cook and some were creations of figures that represented their gods, their fellow human beings or scenes from everyday life. There were decorative pieces. Each people or group left their mark or style that allowed them to be recognised and recognised by other peoples (POINT OF ART, 2016),

In Chã da Pia, ceramics and raw materials have their own expressions, according to Alves (2004). The terms "loiça de barro", "loiça" and "barro de loiça" are used to differentiate themselves from other artisans and similar handicrafts produced in other communities; the term "loiceiro" refers to a very common representation used by the local population: someone who works in the production of "loiças". These terms are local alterations of similar words such as louça, earthenware or louceiro (ALVES, 2004).

Almeida (2010) mentions that "ceramics were made for objective and simple purposes, for everyday use to feed the body and/or the soul". The same author also highlights the issue of gender in the activity; because it was made especially by women, there are few records of men practising ceramics production. Lévi-Strauss (1985) *apud* Almeida (2010) refers to the activity as:

> "The act of making pottery was a sacred, mysterious occupation linked to the feminine universe. Only women who had inherited the right to practise it through other women, their own ancestors. Surrounding the mythology about the origin of ceramics, there were also rigorous methods of harvesting and preparing the clay by women [....]. The Yucararé, a tribe living on the southern edge of the Andes, also surrounded ceramics practices with a series of strict precautions. "The women, who are the only ones to practise this art, went to collect the clay during the period of the year that was not dedicated to harvesting."... the women of the Jivaro tribe were responsible for generating the human, food (planting and harvesting) and ceramic fruits. It was the women who were responsible for planting and harvesting the crops, making and firing the clay, as well as gestating and looking after the children. The men were responsible for all endeavours that required travelling outside the village, such as hunting, fishing, fighting, weapons and anything that required physical strength"

In the Chã da Pia community, the clay pieces are made mainly by women, following the tradition mentioned above and a very common practice in the Northeast, but also present in other regions of the country. Many municipalities in the north-east of Brazil have "louceiros" (potters), including Paraíba and Pernambuco, Rio Grande do Norte, Ceará and Espírito Santo, with the famous black pots from Espírito Santo used to prepare moquecas; Caruaru, Tracunhahém and Goiana stand out for their production of utilitarian and ornamental ceramics. Ceramics can also be found in the municipality of São Gonçalo do Amarante in Rio Grande do Norte and Macaúbas in Bahia (MACHADO, 2016).

Ceramic production takes place all year round in Chã da Pia, but is intensified during the dry season due to the dry climate that benefits production with the use of sunlight and the high temperature that is important for drying the pieces, while in the winter period they take a long time to dry due to the humidity in the environment, and cooking is done outdoors in uncovered areas.

The dry season offers the most favourable conditions for making clay pieces, but it is during this period that the difficulties of accessing water from dams, reservoirs and even sinks appear, if the rains had been scarce in winter. In Chã da Pia, the sinks, the puddles dug naturally in the large flagstones, store a lot of rainwater and it is from these depressions that the artisans draw water to prepare the "crockery clay". Another source they use is the water from ephemeral rivers, but only when the winter is long enough for them to form. When these sources run out, the artisans use good quality water from cisterns distributed by water tankers.

The artisans make use of clay, water and firewood, which are natural resources. They adapt to the conditions of the region, taking the conditions offered by the semi-arid region as a viable strategy to guarantee income and improve their quality of life. Production begins with the search for basic resources, which are found in the environment in which people live. This activity, on a small, medium or large scale, repeatedly becomes predatory.

In many semi-arid regions, the soil contains clay, sand and pebbles. The artisans or potters recognise this and turn to the "canteiras", places with a cleaner mixture of clay and sand, without other components or with them in scarce concentrations, which make it difficult to prepare the clay and finish the pieces.

Few ceramists use pure clay because it shrinks as it dries and breaks (POINT DA ARTE, 2016). For this reason, they prepare a clay paste with sand that makes modelling easier, gives the piece more strength before firing and makes the clay less sticky when wet and less brittle when dry, and more resistant for cooking.

It is the women "loiceiras" themselves who remove the clay from known plots of land located in the community itself or nearby; they need to know the quality of this clay as well as the water, which has a great influence on the final product (fragility of fractures, popcorn on the top layer during cooking, among other problems). In Chã da Pia, in the current drought, the clay is found in the form of dry clods that are broken up with hoes and picks. It is extracted from a neighbouring site with an appropriate soil profile, the B horizons of sodium-affected Planosols, classified as Natic and Haplic Planosols, which are sodic, hypereutrophic, shallow soils with high clay activity (ALVES, 2004).

The earth is transported in hand carts by the artisans or helped by their sons or daughters and placed near the home or where the production takes place. This clay is covered with plastic (figure 11) to prevent it from drying out. The preparation of the paste or dough begins with separating the quantity to be used and placing it in a hole made in the ground, adding water to soften the clods and form a homogeneous mass suitable for modelling. The water used to wet the clay in the current drought comes from cisterns, supplied by water tankers, some with drinking water and others with water from dams, but when it rains (sparse rains) and water accumulates in the sinks, this water is preferred. Water consumption is around 30 litres for the preparation of 5 buckets of clods, which produce an average of 20 pieces. For modelling and finishing the pieces, around 12 litres are used per week. This total of 42 litres is not a huge amount of water, but it is equivalent to the consumption of good quality water by a member of the family for 4 to 5 days, if we consider the expenditure per person per day of 8 to 9 litres as calculated for the PIMC cisterns.

Artisans prefer the water that collects in sinks, stone tanks or clay pits, as it is less saline and more alkaline. Possibly because they are eutrophicated waters with an alkaline pH, they are beneficial for handling and firing. The following pictures show one of the clay removal sites and the stages of preparing the wet clay for modelling (pictures 10 and 11)

Figura 10. (a) Clay at the extraction site; (b) Clay in the yard to prepare the dough for moulding the pieces.

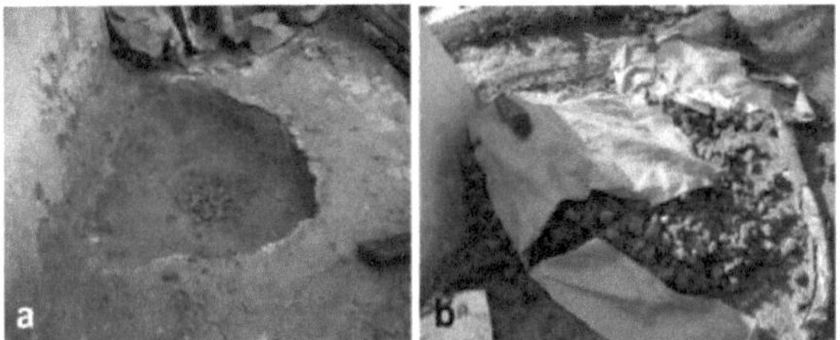

Figure 11. (a) Hole used to prepare the wet clay; (b) Dough moistened and covered with a tarpaulin to prevent drying out.

Once the clay has been moistened, it is homogenised and the pebbles are removed, along with other particles, until a homogeneous and smooth paste is obtained. The addition of sand is practised by many artisans to thicken the clay, as it facilitates modelling and cooking and the pieces break less during firing. Once the mixture has been prepared (water and earth with or without the addition of sand), it is covered with a tarpaulin or plastic wrap for around 24 hours and then the modelling begins.

The technique for making "earthenware" is simple, without the use of a potter's wheel and without any moulds or other tools. It is done completely by hand from a "cake" of wet dough in the shape of an inverted bucket (figure 12). The "potters" work sitting on the floor, using simple tools such as pebbles, animal hide, shards, pieces of wood and metallic blades to shape the pots and polish them. These tools are kept moisturised in a basin of water (figure 13).

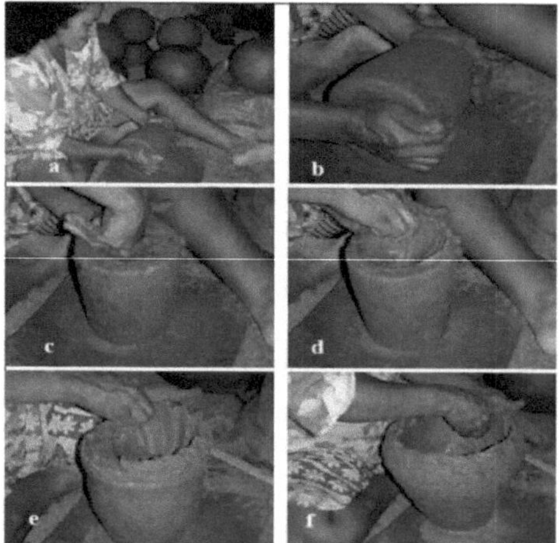

Figura 12. "Loiceira" making a pot in Chã da Pia: (a, b) formation of the clay "cake" (c,d,e,f) transformation of the "cake" into a pot (future pot).
Source: Santos, (2004)

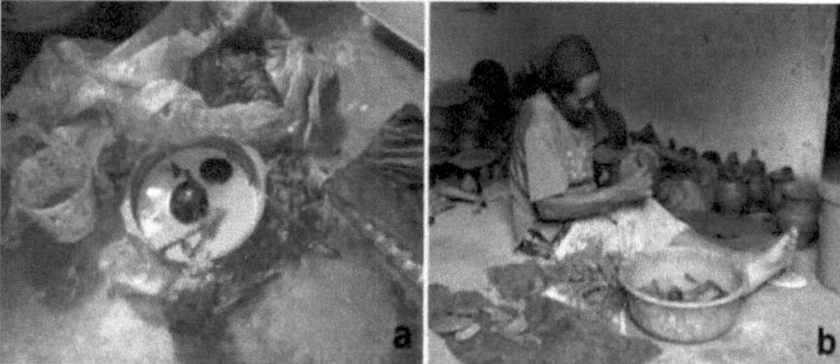

Figure 13. Utensils used to make the clay pieces and to finish them.

The "crockery" is made in their own homes or in sheds belonging to the family. By custom and adaptation, they prefer to work alone, producing their own pieces while taking care of the household activities.

The community has a co-operative which, in 2013, began working to preserve the intangible heritage of the potters with the project "O saber e o fazer das loiceiras da Chã da Pia" through proposal 42485/2012 (FEDERAL GOVERNMENT, 2014). The project began with the participation of a large number of craftswomen, but over time several left the co-operative. Currently there are only four artisans left, who exhibit their pieces (the crockery) at the biggest fairs in the state, such as those in João Pessoa and Campina Grande.

After shaping the "crockery", it dries for a few days at room temperature. This time depends on the weather conditions. After drying and before firing, the final finishing touches are made by removing pebbles and covering any imperfections left on the ceramic surface. It is left to dry again. Cooking takes place in open-air or covered kilns made from the clay itself (figure 14). These have two opposing openings: the lower part houses the fire and the pieces are deposited for cooking in the upper part of the oven.

The firewood used for cooking by "louceiros" is obtained from two plants, one called "amorosa", of the genus *Mimosa* L., and the "marmeleiro", of the genus *Croton.* They produce the appropriate fire for a slow, even burn at a high temperature, which prevents the pieces from bursting or "popping".

Some of these artisans buy firewood at an average price of R$35.00 per square metre, while others collect it themselves. The pieces are organised in the kiln and covered with fragments ("shards") of broken pots, followed by firing (figure 15).

The 10 artisans surveyed use similar production methods, but with slight differences in the finish and variations in the final production line as well as in marketing, so that a hierarchy has emerged between them. By the end of the research it was possible to identify four categories of artisans in relation to the production line: 1) those who produce raw pieces that they sell to other artisans who burn them and resell them; 2) those who prepare the pieces that are sold raw and a small part of which they burn themselves (they have small ones). Some of these pieces are for use in the home, others for direct sale at weekly fairs in neighbouring towns; 3) another group produces the ceramics that are fired by themselves (they have one or more fomos of different sizes) and sold; 4) those who buy raw pieces, fire them and sell them. This hierarchy was born out of differences in production and income, and shows a certain relationship of dependence based on "socio-economic status". The situation of dependency in turn causes social tensions, which make harmonious coexistence difficult and democratic discussions about community actions more difficult.

There are frequent kinship relationships among the artisans: sisters, mothers and daughters, cousins, nieces, sisters-in-law, all dedicated to the same profession. However, the younger generations show no interest in continuing this practice and consider it to have "no future". Young people have the chance to work in other activities, they have access to school and university due to the availability of education credits, and the family has a decent income through the various income transfer mechanisms.

Figure 14. (a) Covered oven; (b) Open-air oven and bundles of firewood.

Figure 15. (a) Potter organising the pieces in the oven (b) Oven filled with the clay pieces and covered with shards of broken pots, ready for the "dishes" to be cooked.

Figure 16. (a) Raw "crockery"; (b) cooked "crockery".

The dishes are sold at fairs in neighbouring towns (Arara, Esperança and Remigio). Prices vary from R$7.00 to R$10.00 for stoves and small pots, and R$20.00 to R$30.00 for larger ones (jugs and large pots). Some "loiceros" take part in craft fairs in Campina Grande-PB (figure 17) and João Pessoa-PB, where prices are higher.

Figure 17. Clay pieces sold at the Campina Grande Crafts Fair in 2013 and 2014 b

5.3. Socio-economic questionnaire

5.3.1 Socio-economic profile of families

Of the 10 artisans interviewed in the Chã da Pia community, nine are women and one is a man. The average age was 55 and the families consist of two to five members. This average figure coincides with that of the rural communities of Malhada da Roça, Poços das Pedras and Curral do Meio, located in the municipality of São João do Cariri (MIRANDA, 2011). This data is important because it confirms ASA's calculations for defining the maximum volume of PI MC cisterns, the area of the roofs and the maximum daily volume for consumption per inhabitant. They are important for the proper management of the systems, especially in the projection that, if the target is reached, there will be one million PI MC systems distributed in the semi-arid region, benefiting 5 million people.

All the interviewees are potters or loiceros (figure 19) and practise family farming (figure 18), selling their surpluses. The average income is R$1,241.00 per family (table 1), but this is an approximate figure because some of the interviewees didn't know precisely what their monthly income was or what their family's income was. The calculation took into account the fact that most of the artisans have a rural pension, a family grant/school grant/youth grant and extra income from ceramics sales, which can be as much as R$600 depending on the month and the fairs attended.

Variables	Media Min. and Max.	(%)
Members - family	4 2-5	nc
Age (years) - child	9 7-11	nc
Age (years) - artisans	55	nc

	46-71	
Occupation	Farmer / Dairy farmer	100
Family income	R$ 1.241 R$ 500- 2.364[1]	nc
Source: Agricolour/vegetables/social of Agro/laundry/aposent. income Agricultural/laundry/aposent/social		60 30 10

Chart 1 - Socio-economic conditions of Chã da Pia artisans. 1: Family with 3 pensioners; nc: does not correspond; Agr.: Agriculture; Aposent.: Retirement; b. social: social grant.

Figure 18. "Loiceiro" in agriculture - rainy season

Figure 19. "Loiceiras" in the production of "loiça" - dry season

In terms of schooling, 70 per cent have not completed primary school, i.e. they left school at the beginning of their basic education. Some are still illiterate (30 per cent), or are functionally illiterate (literate people who sign their name, but don't do full reading or simple maths). The illiterate group is made up of women aged between 56 and 64. According to the IBGE from 2000 to 2010, women continue to have a higher illiteracy rate than men and among these women, the highest illiteracy rates in the country are among older people living in rural areas in the Northeast and North regions. In the Northeast, the illiteracy rate is 19.1%, followed by the North, with 11.2% (IBGE, 2010).

In one "batch" the artisans fire between 50 and 100 pieces of clay, a quantity that varies over the months: less during the rains and more during the droughts, and which also depends on orders from restaurants, participation in regional fairs, sales at more than one open market, etc. With the transfer of income in recent years, economic conditions have improved for these workers, who have had access to new goods, health services, continued studies, a reduction in infant deaths and birth control.

5.3.2 Housing conditions

During the interviews it was possible to observe the families' living conditions and cleaning practices. With regard to the hygiene of the house and yard, the situations observed were classified into four categories: **excellent** when the conditions of cleanliness of the yard were met, household rubbish in an appropriate place and away from the kitchen and cistern, family hygiene, sewage collection in the cesspit without open draining and no animal faeces in the yard; **good** when most of the requirements mentioned were in line with the previous standards; **regular** when half of the items assessed were not in line and **bad** when most of the items mentioned were not met. The results (table 2) showed that 20 per cent of the houses were in excellent hygiene condition, 60 per cent in good condition and 20 per cent in fair condition. None had a bad condition.

Condition	Features
Great	Cleanliness of the yard and house; household rubbish away from the kitchen and cistern; hygiene of family members
Good	When most of the above items have been fulfilled.
Regular	When the majority of items did not agree
Bad	When none of the items are fulfilled.

Table 02 - Hygiene conditions of homes and backyards in the Chã da Pia community.

In addition to the lack of piped drinking water, there is no sewage system and no rubbish collection, i.e. there is no basic sanitation. All the houses are self-contained and have kitchens (some outside, others inside) with fridges, gas and/or wood-burning cookers and water filters, some of which are plastic. All the homes have bathrooms, some with more than one: one inside the house, which is used only for bathing and night-time use, and another outside for general use. As for the presence of a cesspit, eight (80%) of the homes visited have a septic tank in adequate conditions and two (20%) families do not have a cesspit; the waste from the toilets is thrown into the environment and some members defecate in pools, which are also discharged into the environment. The inadequate disposal of human waste, whether faeces or sewage, is the main source of faecal contamination of the environment and the transmission of infectious diseases through water and soil. These infections of viral, bacterial or parasitic origin mainly affect children under the age of five, causing diarrhoea that can lead to dehydration and even death. Some families separate their rubbish; the inorganic materials (paper, plastic, glass and metal) are placed in a hole dug in the ground and burned (Figure 20), while the organic material is used as fertiliser and/or to feed animals. Burning solid waste solves the problem of disposing of this waste, which also transmits infectious diseases. However, intense and frequent open burning affects the environment and contributes to air pollution and the greenhouse effect, as well as being harmful to health. When the fire is not controlled, it can reach large proportions and endanger the biota and many residents who live in the vicinity where the rubbish is being burnt.

Figura 20. Rubbish piling up in a hole in the yard, to be burnt.

5.3.3 Origin and treatment of drinking water

As in many rural communities, in Chã da Pia it is possible to find numerous families who have catchment systems and

storage of rainwater in PIMC cisterns, which guarantees the storage of a significant volume of water for use in times of scarcity. All the households studied have cisterns, some even have two, one from the PIMC and the other built by the family. During the period of long drought when the research was carried out, there was minimal and sparse rainfall, insufficient to store a significant amount of water. It was the water trucks that fed the cisterns, which came to be used as reservoirs for water other than rain, to supply the family.

Some families have organised their own management of the cisterns according to the quality of the water they receive: 1) community and family cisterns in good condition have been selected to receive "drinking water" transported by the pipa operation. This water is used for human consumption (drinking, cooking and bathing); 2) family cisterns that receive good but not potable water or water from a reservoir, supplied free of charge by the town hall and considered to be of "good" quality and used for spending or drinking after being treated in the home if there is no other water of better quality; 3) family cisterns filled with water from private water tankers of unknown origin intended for "spending".

Below is a flowchart of the "path of water" distributed by water tankers in the community (figure 21).

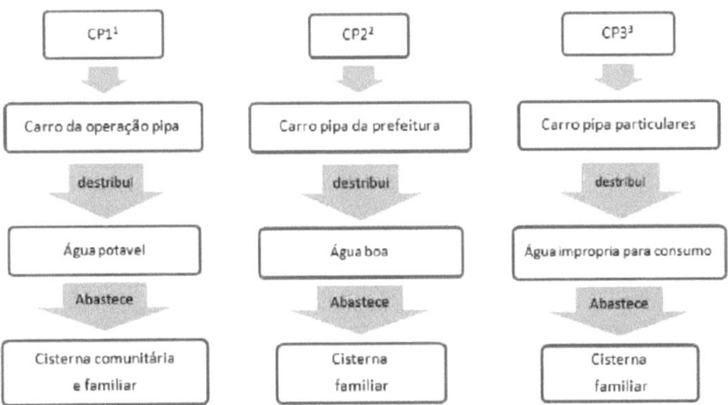

Figura 21. Water route from the water tanker cars to the cisterns in Chã da Pia/Areia/PB. CPI[1]: army water tanker, CP2[2]: municipal water tanker and CP3[2]: private water tanker.

Buying good quality water from a water truck is not viable for many families, who use water from community cisterns filled with water from the water truck operation. These are distributed in strategic places: in the centre of the municipality, in schools and community associations or in homes that they share with nearby families. They contain good quality water and are filled more frequently than family cisterns; they are well looked after and their management follows hygiene principles. Of the 10 participating families, 70% have only one cistern, either from the P1MC or their own (30% are their own and 40% are from the P1MC), and 30% have more than one cistern (Own and P1MC), as shown in the following graph:

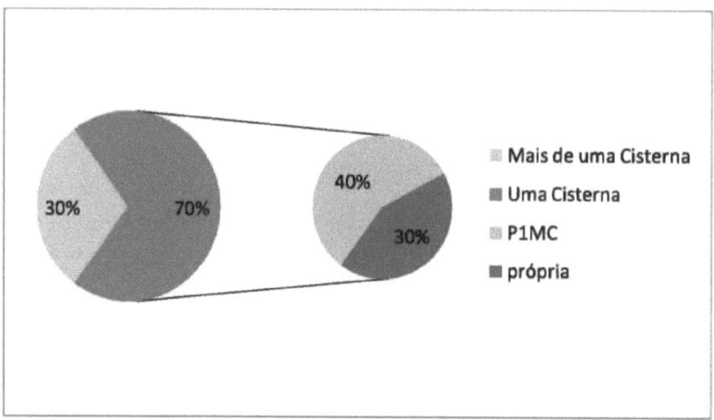

Graph 01. Percentages of cisterns distributed per family in the Chã da Pia community - Areia/PB

With no or insufficient rainfall, the community has adopted alternative water management, realising the importance of separating water that is good for consumption from water that is wasteful. This behaviour responds to the community's ancestral knowledge and perceptions, repeating the actions of their parents and grandparents and the teachings and actions of the ASA- PI MC- SUS. The little rainwater that accumulates in the sinks and clay pits is used for "waste", in the production of "crockery" and for watering small animals (figure 22).

The long drought affects the volume and management of the water stored in the cisterns, which cannot be filled and thus be able to satisfy the greater demand, which grows with the prolongation of the drought, and especially as the dams, the main source of water in the semi-arid region, dry up. In extreme situations, this poor quality water is used by some families for all purposes when the scarce sources dry up completely (sinks and dams) and there are no others. Some families filter this inferior quality water before using it for cooking and drinking, but it has not been reported that they disinfect it.

Figure 22. (a) Barreiro used as a source of water during the dry season, (b) Rainwater accumulated in a large sinkhole on a slab in Chã da Pia, generally used to make earthenware.

Most of the PI MC cisterns in the community were built in a joint effort within the context of PI MC, while others were built by the families themselves (figure 23). All of them have a mobile gutter system to eliminate the first waters of each rainfall event that drain from the catchment area (the roofs), i.e. when the downpipes are removed, the first waters, which are the dirty waters from washing the roof, are allowed to drain. To avoid having to manually remove and replace the downpipes, you can include a device for diverting the first rainwater, as recommended by the ASA, between the downpipe and the cistern. These are small plastic or brick and mortar boxes with a volume of approximately 60 to 80 litres for roofs of 50 to 80 metres (figure 24). When calculating the capacity that these water diversion systems should have, it should be remembered that 1 mm of rainwater falling on one metre produces 1 L of water. These diversions accumulate the dirty water that runs off the roofs and, when full, allow the cleaner water to overflow into the cistern (XAVIER et. al, 2012). The cistern with the stored dirty water must be emptied after each rainfall event and can be used to irrigate the fields and to feed the animals.

Figure 23: (a) Cistern built by PI MC; (b) Cistern built by residents (both without the system for diverting the first rainwater).

Figure 24. Cisterns with systems for diverting the first waters.
Source: Xavier, 2012

Around 90 per cent of the families said that the rains in the region (when they are within the historical average) are enough to fill the cistern, ensuring drinking water until the following year. However, in the drought of the last four years, the water collected is not enough to satisfy a family of five for three to four months.

Just as in Chã da Pia, throughout the Brazilian semi-arid region women used to carry out the task of supplying their homes with water, having to walk several hours a day to the source and return carrying cans of water. All of the participating artisans/farmers mentioned that after the cistern was built, they now had access to water in their backyards that was of a higher quality than traditional sources (dams, cisterns, rivers, reservoirs, water trucks). The ease of having good water close to home without having to walk long distances every day or even more than once a day to fetch water of dubious quality from cisterns, dams or reservoirs has substantially improved the daily lives of families and their wives. As a result, families' health conditions improved, with a reduction in diarrhoeal diseases. The women were freed from these long walks and the time they gained was and is used to restart their studies, help the children with schoolwork, take part in community activities, among other activities mentioned in some of the testimonies recorded in the field and presented below:

> *" ... before the cistern we used to get water from the tanks...after the cistern was built it improved, it improved a lot because before we didn't have any water, right, now whenever you want to fetch water you have it in the cistern...in the old days when there was no cistern, when it rained we would go and fetch it from the slabs, down there, and if it rained at dawn no one would sleep anymore, everyone would wake up to fill the pots, buckets, jars...whoever fetched it the quickest would fill it up...whoever slept too long would run out of water."*

> *"Today those who don't have a cistern here in Chã da Pia suffer...after we built the cistern, it's a blessing from Jesus...because when we're without water, God sends the rain, or buys the water, or goes to the community to get it. After the cistern I stopped fetching water far away, I used to fetch water more than once a day".*

The cisterns have brought and continue to bring improvements to families in the semi-arid region, by providing good quality water free of charge (in the rainy season) and freeing them from having to ask for water from the owners of the land and dams, restoring self-esteem and stimulating the use of citizenship.

5.3.4. Management and conservation of rainwater harvesting and storage systems during droughts.

The rainwater collection system must be properly maintained, as must the cistern, and it is the families who carry out this task, as well as the handling of rainwater, from collection to storage and final treatment before consumption (disinfection). This care must also be maintained during the dry season, in order to receive and preserve the water from the water trucks.

With regard to cleaning the catchment area, 70 per cent of those interviewed said that they only clean the gutters and downpipes when it rains and 30 per cent do not take the necessary care. Various types of dirt were found on the roofs, such as leaves and various types of rubbish. Of the inadequacies presented, the most frequent in the cisterns were broken or bent lids, and irregularly positioned gutters and downspouts in the collection system (figure 25). One of the interviewees commented on the positioning of her "spout":

> *"...my cistern didn't fill up very much, you see! Because of the small spout, there are times when it rains and it doesn't fill up... it doesn't even last the whole year .... my mum's with this rainfall is almost full, but mine doesn't fill up every winter, because the spout is small, you have to adjust it and put it on the other side too "*

In this speech, it was clear that the importance of the positioning of the gutter and its size was recognised, which prevented more water from being collected from the roof, which would have increased the volume to be stored in the cistern. Figure 25 shows some of the inadequacies mentioned.

Figure 25. Irregularities in the catchment systems of families in Chã da Pia. A) Bent gutter; b) Gutter outside the drainage area; c) Gutter with accumulated dirt.

Cisterns should be cleaned at least once a year, using brushes, cloths, bleach, good quality water for the final rinse and a squeegee; cracks should be covered with cement and when dry, they should be painted with lime on the outside . Maintenance work was carried out on some cisterns (sealing cracks, changing lids, repairing pumps) and a few remained with cracks and/or infiltration. The inadequacies recorded are summarised in table 03 below.

Variable	Results
State of repair of the cistern	Regular/broken lids 70%; good 30%
Inadequacies in SCAAC	Misaligned/dirty gutters/ducts 60 per cent; good 40 per cent
Clean the cistern/frequency	Yes 100% (when dry 50%; once a year 40 per cent; once in two years 10%)
How cleaning is done	Bleach/brush/cloth/squeegee 100%
How to draw water from cisterns	Bucket with rope 50%; cans with rope 50%, pumps 0%
Toilets (type and quantity)/location	w/2 bathrooms 40% (one outside and one inside the house 20%; 2 inside 10%; 2 outside 10%); w/bathroom 50% (inside 10%; outside 40%); 2 indoors and 1 foral 0%

Table 03. Housing conditions and sanitary aspects of the families interviewed in the Chã da Pia/Areia/PB community.

The analyses of the *on-site* observations showed that the rainwater collection and storage systems (when it rains) are irregularly maintained, with 60 % of the systems in poor condition, which makes it difficult to collect the water that runs down the roof efficiently in a rainy event. The family needs this volume to reduce their water vulnerability and get through the dry season without suffering. Mismanagement of the system contributes to families looking for old sources and continuing to rely on water from water trucks, fuelling the situation of dependence on the owners of the reservoirs and remaining prey to regional political power and emergency actions.

Some of the interviews revealed the use of fish (piabas) in the cistern to indicate the quality of the water, a very common and widely used practice

by families because they believe that fish improve water quality by feeding on larvae and suspended materials, and also to reduce the incidence of the *Aedes aegypti* mosquito, preventing dengue outbreaks. This act is seen as a cultural habit. Families take hygienic care with their drinking water, but there are some "loopholes" in the management caused by ancestral habits such as using a bucket to remove water from the cistern that facilitate microbial contamination. All the cistern owners in Chã da Pia use buckets and zinc cans tied to ropes to remove the water. They don't use the hand pump, which is the proper and hygienic practice to avoid contamination. The pump should be used to prevent the cistern lid from opening and the water from coming into contact with the bucket/can, which is left on the ground or out in the open with the rope, which is not washed. Many claim that the pumps are fragile and that the pumping process is tiring when done several times during the day due to the small volume withdrawn

Andrade Neto (2013) points out that direct contact between people and utensils (buckets, cans and ropes) with water can cause cross-contamination, so animals should not have access to water, nor should they be near accumulations of other debris, dust or contaminated water. The incidence of light on the water also affects its quality, as it facilitates the proliferation of algae and cyanobacteria, making the water unfit for human consumption.

Disinfection is the most effective technique for guaranteeing the microbiological quality of drinking water. Ordinance 2914/2011- MS requires a residual of between 0.5 and 2 $mg.L^{-1}$. Of the families interviewed, less than half (4) said that they do not treat their drinking water at all and more than half (6) said that they use sodium hypochlorite in the pot or filter where their water is stored. Graph 02 shows the percentage of families who said they chlorinate their drinking water.

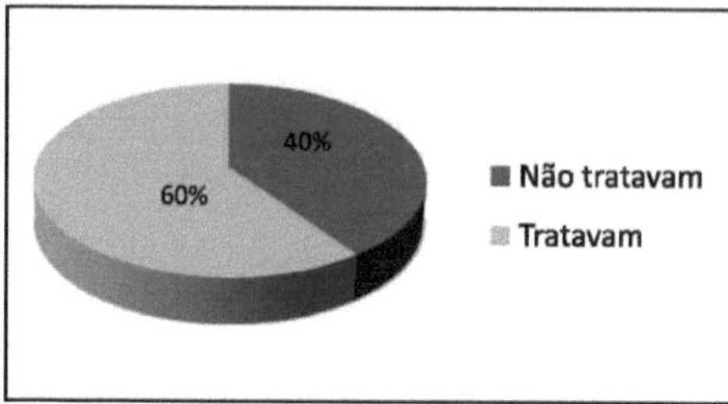

Graph 02 - Percentage of families that say they chlorinate their water.

But in practice there are some differences, because this is a ready-made discourse that doesn't portray everyday life: during the interviews, contradictions were revealed by those who claimed to carry out chlorination. Among the statements, the following stands out:

> *...I put three or four drops in the pot...because it's bitter.*
>
> *...I only add 3 drops because the water is very clean, so you don't need them.*

It is noteworthy that the number of drops of sodium hypochlorite to be added to approximately 20 litre jars and 50 litre jugs does not correspond to what is appropriate, demonstrating that the interviewees do not know how to use hypochlorite correctly (or how to calculate it) to disinfect water before consumption. The volumes of sodium hypochlorite used do not correspond to the recommendations of the SUS, which are written on the label of the bottle that is distributed free of charge by the community health agents (ACS). The instructions say that for every litre of water for human consumption, two drops of 2.5% sodium hypochlorite should be added and left to stand for 30 minutes before drinking (BRASIL, 2015). Many families who don't disinfect their water claim that it comes "pure", "clean" and "cured" and that they don't need to add anything to it. This discourse is transmitted by the army through the pipa operation. Here are some testimonies about the use of chlorine

> *"...the water is very blue, I don't add anything because it's already cured".*

"...nobody gets ill here at home... I've never had to put anything in, and in that time I've brought up my children all drinking water from the reservoir, they've never got ill."

According to those interviewed in the community, visits by community health agents are quick and guidance is only given when necessary. When asked about the distribution of sodium hypochlorite by the health agents, everyone said that they receive bottles of sodium hypochlorite. However, when asked for the containers, several answers were given to justify their absence, which raises doubts as to whether the CHWs actually distribute them with the frequency established by the Department of Health and whether they correctly advise users on how to efficiently disinfect the water before consumption. One interviewee's account of the distribution of sodium hypochlorite is reproduced below:

"The chlorine is brought by the health agent, but she hasn't brought it for over two years, so I put bleach on it...but then last week she brought it...the health agent doesn't come by much...she comes once a month...she doesn't say anything about it, she just talks about my diabetes treatment... "

It was found that the community does not distribute sodium hypochlorite at the monthly frequency stated, and that there is no control of the quality of the water consumed by families. From an epidemiological point of view, it is important to know the quality of water intended for human consumption, which must comply with the potability standard (Ordinance 2914/2011- MS). The ordinance that establishes the quality standards for drinking water for human consumption does not refer to the quality of the water in the cisterns, nor does it establish any monitoring frequency. This is the responsibility of the user, who must take care of the quality of the water to be consumed by the whole family, and which implicitly must be good water in order to comply with the aforementioned ordinance. In this context, SUS Community Health Agents and Environmental Health Surveillance Agents should encourage families in the diffuse communities of the semi-arid region to practise chlorination at the final point of consumption in accordance with the appropriate technical standards to maintain a free chlorine residual of no less than 0.2 to 2 $mg.L^{-1}$. These conditions guarantee microbiologically safe water for human consumption.

The members of these communities and their families must always manage the water correctly before consumption. Families must realise the value of water quality for their health and of the entire water collection and storage system, not just the cistern. They need to understand how it works and consider the importance of having their own good quality water in their backyard.

5.3.5. Clay production and its impact on the environment

The manufacture of ceramic pieces in the community depends on the availability of clay, water and firewood. These materials are present in nature at no cost: the plants are indigenous and grow around the community; some people also have plantations of mimosa and quince, which after cutting are prepared into bundles that are sold to the "loiceiras". The appropriate clay is also found in neighbouring areas, and in the community of Chã da Pia around 40 "loiceiros" obtain clay from the same place, Seu Miguel's farm. An *on-site* visit to this soil extraction site revealed numerous holes and depressions dug in the ground (figure 26), due to extraction since the profile of interest is located approximately 40 - 50 cm from the surface (the "loiça" clay). The soil is removed preferably from areas close to the roots of the local vegetation, as these are wetter and easier to dig. This exposes the roots of the plants, which can lead to their death.

The actions taken by artisans on the soil and vegetation modify the landscape and alter the region's biodiversity. The National Plan to Combat Desertification (PNCD) has analysed that much of the land that is subject to desertification is located in the semi-arid and sub-humid areas of the Northeast. Trees protect the soil against erosion and the action of rain by slowing down run-off and preventing or reducing the dragging away of fragmented soil; without the protection of tree roots, the soil is vulnerable, its impoverishment is favoured and it becomes unsuitable for productive crops.

Figure 26. (a) Soil eroded by the removal of clay; (b) exposed tree roots.

The "clay pottery" is removed once a week or sometimes even monthly, depending on the artisans' production rhythm. During the interviews, the artisans did not mention the possible negative effects on the environment of their actions on the soil, vegetation and water. When asked about the use of these resources, some said that they didn't cause any damage to the environment, or simply "didn't know". It can be considered that they don't realise the changes caused to the soil and vegetation, which are not seen as predatory actions, but rather as a guarantee of a monthly income, mainly for women who achieve a certain financial dependence on the pottery activity. For Cavalcanti (2011), the result of these actions has been an intense process of socio-environmental degradation in the region, "either by those who have access to and concentrate natural resources, or by those who pressurise them in the little or no space they have, as a survival strategy".

It wasn't possible to visit the places where they harvest the mimosa and quince trees to produce firewood for cooking their "crockery", because most of the artisans buy from third parties. The use of firewood for cooking is intense, as some do it every week and buy it frequently. Deforestation leads to the exhaustion of the soil, rising temperatures, the extinction of species and an increase in the greenhouse effect. These are, among other things, unfavourable consequences for all forms of life (NOBRE, et al, 2007)

Figure 27. Vegetation (firewood) used to cook the dishes

These practices are part of the day-to-day life of the loiceiros' families, but they can be used in a sustainable and non-predatory way. According to Miranda (2011) "rural communities have a cultural structure that needs to be known, assessed and respected before any education project, environmental or otherwise, can be introduced". This means that although there are attempts, realisations and projects to change perceptions, there is a whole history of generations and cultures involved. It is essential that managers and government officials take a fresh look at the real needs of these families, who need actions that guarantee coexistence with the semi-arid region without encouraging its depredation. Users' awareness must be reflected in the care they take of their environment, seeking to achieve conditions of sustainability rather than aggression.

CHAPTER 6

FINAL CONSIDERATIONS

Man's coexistence with the semi-arid region is not only feasible, it is a fact shown day by day by the communities that have developed, are developing and remain there.

The lack of rainwater during drought periods increases the search for alternative sources of water, with water tankers (from Operação Pipa, municipal and private water tankers) distributing the largest quantities of water and reaching even the most distant places, but the different origins of these waters suggest associated risks, which must also be taken into account in terms of transport, the hygiene of the equipment used and the handling of this water by families from the reservoir (the cistern) and in their homes.

It is important to insist on improvements in sanitary (hygienic) conditions in the management of water consumed by families, which require actions that must be incorporated into the daily routine, and which require greater control when the water distributed is from an unknown source or there are no guarantees that it will maintain the quality of its origin, as in the case of drinking water from the Pipa Operation.

Environmental education actions, diversified technological alternatives for water storage and incentives for complementary family farming activities are important in the process of social and economic improvements.

The multiple waters used in the community could be better utilised with other technologies such as underground dams and sewage treatment, considering the feasibility of their safe reuse in family farming, technologies that should not be rejected.

Educational actions to preserve the environment and on the use of these technologies that facilitate water availability and reduce water vulnerability must include the active participation of artisans in decision-making processes about the resources they use. Artisan communities such as those in Chã da Pia are generally marginalised. It's important to consider formal avenues for integration and greater visualisation of their work. New public policy practices could be implemented in the community for the cultural preservation of their craft techniques, under more dignified and favourable conditions.

REFERENCE BIBLIOGRAPHY

AESA - EXECUTIVE AGENCY FOR WATER MANAGEMENT IN THE STATE OF PARAÍBA. Available at: http://site2.aesa.pb.gov.br/aesa/volumesAcudes.do?metodo=preparaUltimosVolumesPorMunicipio Accessed 21 Sep. 2016.

ALVES, A. G. From *"earthenware clay" to "earthenware crockery": Ethnopedological characterisation of a peasant craft in the agreste of Paraíba.* 2004. 179f. Thesis (Doctorate in Ecology and Natural Resources) - Federal University of São Carlos, 2004.

ALBUQUERQUE, J. P.T; RÊGO, J.C.; The Brazilian semi-arid region: general aspects. In: GALVÃO, C.O et al. (Orgs.). *Water resources for coexistence with the semi-arid region: approaches by researchers in Brazil, Portugal, Cape Verde, the United States and Argentina.* Porto Alegre: ABRH; Recife: Ed. Universidade da UFPE, 2011. P 19-48.

ALMEIDA;. F. L. *In the beginning was ceramics: a return to the origins.* São Paulo: Editora UNESP; São Paulo: Cultura Acadêmica, 2010. 238 p.

ALMEIDA, DE A. et al. All about the water crisis. Folha de São Paulo. São Paulo, September 2014. Available at: <> http://arte.folha.uol.com.br/ambiente/2014/09/15/crise-da-agua/index.htmAccessed 22 Jun. 2016.

ANDRADE, J. A.; NUNES, M. A. Access to water in the Brazilian semi-arid region: an analysis of public policies implemented in the region. *Revista espinhaço,* v.3, p 28-39, 2014.

ANDRADE NETO, C. O. Immediate Utilisation of Rainwater. Revista *Eletrônica de Gestão e Tecnologias Ambientais (GESTA).* v.l, n.l, 2013, p. 073086.

ARAÚJO, J. C. Water management of small reservoirs in the semi-arid region. In: APHA - AMERICAN PUBLIC HEALTH ASSOCIATION. *Standard methods for the examination of water and wastewater.* APHA, AWWA, WEF, 2012. 1946 pag.

ASA BRASIL. Semi-arid. Available at:< http://www.asabrasil.org.br>. Accessed on: 25 Apr. 2015.

ASA BRASIL. Semi-arid. Available at:< http://www.asabrasil.org.br>. Accessed on: 18 July 2016.

ASA BRASIL. Semi-arid. Available at:< http://www.asabrasil.org.br>. Accessed on: 22 Mar. 2017

BAVA, S. A. C.(2004) Social technology and local development. In: PAULO, de A. et al. (Eds). *Social technology: a strategy for development.* Rio de Janeiro: Banco do Brasil Foundation, 2004. P. 103-116.

BRAZIL, 2005. New delimitation of the Brazilian semi-arid region. Ministry of National Integration, 33p.

BRAZIL 2011. MINISTRY OF HEALTH, ORDINANCE 2914. Available at: http://bvsms.saude.gov.br/bvs/saudelegis/gm/201 l/prt2914_12_12_2011 .html> Accessed on 11 Dec 2015.

BRAZIL. Care with water for human consumption. Available at: http://bvsms.saude.gov.br/bvs/folder/cuidados_agua_consumo_humano_2011 ,p df> Accessed on 28 Nov. 2015.

BRAZIL. Water crisis worsens in the Brazilian semi-arid region. Available at: http://www.brasil.gov.br/meio-ambiente/2016/09/crise-hidrica-se-agrava-no- semiarido-brasileiro Accessed on: 30 Sep. 2016.

CAMPOS, J. N. B. Droughts and public policies in the semi-arid region: ideas, thinkers and periods. *Estududos avançados,* São Paulo , v. 28, n. 82, 2014, p. 65-88.

CAVALCANTI, E. R. Environmental education and contextualised education based on coexistence with the semi-arid region. In: LIMA, R. da C. C. et al. (Ed.). *Desertification and climate change in the Brazilian semi-arid region.* Campina Grande: INSA-PB; 2011. P 79-88.

CEBALLOS, B. S. O.; DANIEL L. A,; BASTOS, R. K. X. Water treatment for human consumption: world panorama and Prosab actions. In: PÁDUA, V. L. (Org). *Removal of emerging microorganisms and organic microcontaminants in the treatment of water for human consumption.* Rio de Janeiro: ABES, 2009. P. 19-43.

CIRILO, J. A. The Brazilian Semi-Arid: Water Policies. In: GALVÃO, C. O. et al. (Orgs.) *Recursos Hídricos para a convivência com o semiárido.* Porto Alegre: ABRH; Recife: Ed Universidade da UFPE, 2011.

COSTA. A. B.; DIAS. R. de B. State and civil society in the implementation of cistern policies. In: COSTA. A. B (Org). *Social Technology and Public Policies.* São Paulo: Pólis Institute, 2013. p. 33-63.

DAGNINO, R. et al. On the analytical-conceptual framework of social technology. In: Banco do Brasil Foundation (Org). *Social technology: a strategy for development.* Rio de Janeiro: CIP-Brasil, 2004. P. 15-64.

Dl BERNARDO , L.; DANTAS, A. *Methods and Techniques of Water Treatment.* Ed. RIMA, 2005, pl565.

DNOCS - National Department of Works Against Droughts. Available at <http://www.dnocs.gov.br/php/comunicacao/registros.php?f_registro=2&> Accessed on 12 May, 2016.

FBB - FUNDAÇÃO BANCO DO BRASIL - Satiable technologies. Available at:< http://www.fbb.org.br/tecnologiasocial/o-que-e/tecnologia-social/>. Accessed on: 08 May 2015.

FERREIRA, L. F. G.; Drought as a political and social manifestation: Oligarchies and Cangaço in Paraíba. *History of Paraíba,* Nov 2008. Available at: <

http://historiadaparaiba.blogspot.com.br/2008/ll/seca-como-manisfestao- politico-social.html> Accessed 09 Sep. 2016.

GNADLINGER, J. Rainwater harvesting: A tool for serving rural populations in arid and semi-arid locations. In: MEDEIROS, S. S. et al. (Eds). *Water resources in semi-arid regions.* Campina Grande: INSA, 2011. p.325-360.

FEDERAL GOVERNMENT - 2014. Available at: http://www.portaltransparencia.gov.br/convenios/DetalhaConvenio.asp7CodConvenio=774938&TipoConsulta=0&UF=PB&CodMunicipio=1921&CodOrgao=&Pagina=&Periodo=. Accessed on 18 May 2015.

GUALDANI, C. et al. Integrating Experiences and Reapplying Knowledge. In: GUALDANI, C. et al. *Convivência Com Semiárido Brasileiro". Reapplying Knowledge Through Social Technologies.* Brasília: LABS - Brazil, 2015. p. 15-40

HEIJNEN, H. Rainwater harvesting: water quality, health and hygiene aspects. In: SANTOS, D. B et. al. (Org.). *Rainwater Harvesting, Management and Use.* Campina Grande: INSA, 2015. P. 77-90.

IBGE, 2010 - Brazilian Institute of Geography and Statistics. Available at: < http://censo2010.ibge.gov.br/resultados.html> Accessed on: 22. Aug. 2015.

IBGE. 2015. Brazil has more than 204 million inhabitants, says IBGE . UOL NOTICIAS. Available at:< http://noticias.uol.com.br/cotidiano/ultimas- noticias/2015/08/28/brasil-tem-mais-de-204-milhoes-de-habitantes-diz- ibge.htm.> Accessed on: 07 Sep. 2016.

JÚNIOR, A. E. L.; PEDREIRA, J.S.; Social technologies and public policies. In: LIMA, P. R. L.; Technological aspects of slab cisterns. In: SANTOS, D. B.; et al. (Orgs.). *Rainwater harvesting, management and use.* Campina Grade, PB: Instituto Nacional do Semiárido, 2015 p 341 - 354.
LÉVI -STRAUSS, C. A *poteira aumenta-*, translated by José António Braga Femandes Dias. Lisbon: Edições 70, 1985.

LUCCHESI, F.; Works against droughts and the internalisation of bureaucracy: the action of DNOCS in the hinterland of Paraíba. *Instituto de Investigaciones Antropológicas de Castilla y León (IIACyL).* n 2, p. 51-59, jun. 2016.

LIMA, P. R. L.; Technological aspects of slab cisterns. In: SANTOS, D. B.; et al. (eds.). *Rainwater harvesting, management and use.* Campina Grade, PB: Instituto Nacional do Semiárido, 2015 p 341 - 354.

MACHADO, R. C. V. Clay handicrafts. Pesquisa Escolar Online, Joaquim Nabuco Foundation, Recife. Available at: <http://basilio.fundaj.gov.br/pesquisaescolar>. Accessed on: 11 March 2016.

MALVEZZI, R. *Semiárido: uma visão holística.* Brasília: Confea, 2007. p 140.

MARENGO, J. A. et al. Climate Variability and Change in the Brazilian Semi-Arid. In: MEDEIROS, S.; et al. (Orgs.). *Water resources in arid and semi-arid regions.* Campina Grande, PB: Instituto Nacional do Semiárido, 2011p383-422.

MINISTRY OF THE ENVIRONMENT. Available at< http ://www.mma. gov.br/estruturas/secex_consumo/_arquivos/3%20-%20mcs_agua.pdfhttp:/www.mma.gov.br/estruturas/secex_consumo/_arquivos/ 3%20-%20mcs_agua.pdfl> Accessed 05 Apr.2016

MIRANDA, P. C. Cisternas *no cariri paraibano: avaliação das práticas de educação ambiental no uso higiênico da água.* 2011. 96f. Dissertation (Master's Degree in Environmental Science and Technology) - State University of Paraíba, Campina Grande, PB, 2011.

NOBRE, C. et al. Climate Change and the Amazon. *Science and Culture.* São Paulo: vol.59 n°3, Jul/Sep. 2007. Available at

<http://cienciaecultura.bvs.br/scielo.php7picks0009-67252007000300012&script=sci_arttext>. Accessed on: 23 Mar. 2016.

OLIVEIRA, L.L. The creation of Sudene. Centre for Research and Documentation of Contemporary Brazilian History (CPDOC). São Paulo, 2015. Available at: https://cpdoc.fgv.br/producao/dossies/JK/artigos/Economia/Sudene Accessed on 24 Jan 2016.

ONUBR. (2015). By 2050, one billion people will live in cities without enough water, says World Bank. Available at: <https://nacoesunidas.org/ate-2050-um-bilhao-de-pessoas-viverao-em-cidades- no-water-sufficient-says-world-bank/> Accessed 05 Sep. 2016.

PAND - 2012. National Household Sample Survey. Available at: http://www.ibge.gov.br/home/estatistica/pesquisas/pesquisa_resultados.php7id_ pesquisa=40 Accessed on 13/08/2015.

PHILIPPI JR, A,; MARTINS. G. *Águas de abastecimento e Saneamento, Saúde e Ambiente: Fundamentos para um desenvolvimento sustentável.* Ed. Manole, 2005.

PIELOU, E. C. *Freshwater.* Chicago: The University of Chicago Press, 1998.

POMPONET. A. S. 100 anos de DNOCS: marchas e contraramarchas da convivência com as secas. *Conjunto & Planejamento,* Salvador, n.162, 2009, p.58-65.

POINT OF ART. History of ceramics. Available at: < http://pointdaarte.webnode.com.br/news/historia%20da%20cer%C3%A2mica/> Accessed on 05 Mar. 2016.

PORTO, R.E. et.al; Conservation and rational use of water in rain-dependent agriculture. In: MEDEIROS et.al (Ed). *Water resources in arid and semi-arid regions.* Campina Grande: INSA, 2011. p. 59-84.

REBOUÇAS, A. DA C. Water and rural development. *Estudos Avançados,* São Paulo, vol.15 n°.43 Sep./Dec. 2001. Available at:< http://www.scielo.br/> Accessed on: 14 Nov. 2014.

REBOUÇAS, A. DA C. Water in the Northeast: waste and scarcity. *Estudos Avançados,* São Paulo, vol.ll n°.29 Jan/Apr. 1997. Available at:< http://www.scielo.br/> Accessed on: 14 Nov. 2014.

SANTOS, A.C. *Limits and possibilities of participation in the one million cisterns programme.* 2012. 153f. Dissertation (Master's in Regional Development) - State University of Paraíba, Campina Grande, PB, 2012.

SANTOS, A. C.; CEBALLOS, B.S.O.; SOUSA, C. M. Public water policies and participation in the semi-arid region: limits and tensions in PI MC. *GESTA 1* : p. 139-145, 2013.

SILVA, R.M.A. Entre dois Paradigmas: combate à seca e convivência com o semiárido. *Sociedade e Estado,* Brasília, vol.18, n°.l-2, Jan/Dec. 2003.

TAVARES, M.C et.al; SECA E PODER: ENTREVISTA COM CELSO FURTADO. Ed: Perseu Abramo Foundation. São Paulo, 1998, 83p.

TUNDISI, J.G.; MATSUMURA-TUNDISI, T. *Recursos Hídricos no século XXI.* Ed:

Oficina de Textos. São Paulo, 2011, 328 p.

TUNDISI, J. G.; *Waters in the 21st century.* São Carlos: Rima, IIE, 2, Ed., 2005. 248p.

VIEIRA, A. DE R. Mirror of water . In: VIEIRA, A. DE R. *Agua para vida, água para todos.* Brasília: WWF-Brazil, 2006. P. 8-15.

XAVIER, R.P.; CEBALLOS, B. S. O.; NOBREGA, R.L, B.; GALVÃO, C.O. Analysing the influence of sanitary barriers on the quality of rainwater *stored in rural cisterns.* In: 11th Northeast Water Resources Symposium. November/2012/João Pessoa.

Annexes

SOCIO-ECONOMIC QUESTIONNAIRE

1. GENERAL INFORMATION

Municipality:	District:	Neighbourhood:	Location:
Date of Report:	(/)	/ Hour (h):	Photos: by_____ a_____
Address Domicile: Interviewer			N°
GPS localisation	Latitude:	Longitude:	Altitude:

♦ first letter of interviewer's name / questionnaire no.

2. HOUSEHOLD QUESTIONNAIRE

II. CHARACTERISATION OF THE INTERVIEWEE AND RESIDENTS
NAME: ____________________________________ **SEX:1.M** 2.F 2.1. Schooling: ____________________ 2.2. Occupation ______________ 2.3. Income______________________ 2.4. Do people live in the house? _________ 2.5. Children under five? ______________________

2.6. Children over five? ______________________________

2.7 Have you always lived here?

1. YES 2. NO

2.8. Current property condition

1. Own 2. Owned on loan 3. Owned 4. Other

2.9 Duration of the rainy season

1.4 months 2. 6 months 3. 8 months 4. Don't know

2.10 When does it rain in which months?

2.11 In which month does it stop raining?

2.11. Does it rain a lot during these months?

III. DESCRIPTION OF WET AREAS (KITCHEN)

3.1 Does it have a kitchen?

1. YES 2. NO

3.2 The kitchen has

1. Sink with continuous water 2. Sink without continuous water

3. Gas cooker 4. Wood-burning cooker 5. Improvised cooker

6. Fridge 7. Filter

3.3 Kitchen location

1. Outside the house 2. Inside the house.

VI. WATER SUPPLY

4.1 Does your home have running water?

1. YES 2. NO (Skip to question /4.3)

4.1.1 Is the piped water treated?

1. YES 2. NO

4.2. Storing tap water

1. Water tank 2. Cistern 3. Pot 4. Filter 5. Tonne

6. Other:______________

4.3 Where does your family's drinking water come from?

1.Well 2. Spring 3. Cistern 4. Fountain

5. River 6. Lake 7. Dam/weir 8. Other

4.4. Where does the water come from that your family uses for personal and household hygiene?

1 Well 2. spring 3. Cistern 4.

Fountain

5. River 6. Lake 7. Dam/weir

8. Other: ______________

4.5. Drinking water storage

1. Water tank 2. Cistern 3. Pot 4. Filter 5. Tonne

6. Other:

4.6. Storing water for personal and household hygiene

1. Water tank 2. Cistern 3. Pot 4. Tonne

5. Other: __________

4.7. Does the family treat the drinking water?

1. YES 2. NO (Skip to question 4.10) 3. Coming treated (Skip to question 4.10)

4.8 Ways of treating drinking water (multiple answers)

1. Filtration 2. Chlorination 3. Boiling 4. No treatment

5 .Other: __________

4.9 Material used for filtration

1. Clay filter with candle 2. Charcoal/sand filter

3 .Other_____________

4.10. How do you chlorinate water?

1. YES. What is the dosage?______________

2. No (Skip to question 4.10) 3. Does not use chlorine (Skip to question 4.10)

4.11. Chlorine (sodium hypochlorite) is added

1. In the filter 2. In the cistern 3. In the water tank 4. In the pot

5. No chlorine

4.12. Do you put rainwater in the cistern when it rains?

4.13. How long have you been using rainwater stored in a cistern?

1.<lano 2. 1 year 3. 2 years 4. 3 years 5. 4 years

6. 5 years 7. > 5 years

4.14. The water from the cistern is used for (multiple answers)
1. Drink 2. Cooking 3. Bathing 4. Cleaning 5. Washing clothes 6. Irrigation 7. Other: ___________
4.15. The water in the cistern is of good quality
1. YES 2. NO
4.16. Observed aspects:
1. transparent 2. Cloudy 3. Suspended material 4. Has colour 5. Has an odour 6. Has a flavour 7. bugs in the cistern 8. Type of bugs ______________
4.17. Origins of the cistern
1. Built by the resident 2. Social project (P1MC) 3. Association 4.Other: ___________
4.18 Type of use of water from the cistern
1. Family 2. Collective
4.19. Cistern model
1. Plates 2. Ferro-cement 3. Masonry 4. Don't know 5. Other:____________
4.20. What was your family's water supply like before the cistern?
1.Well 2. Spring 3. Fountain 3. River 4. dam/weir/pond 6. Other _______________________
4.21. Has the installation of the cistern brought improvements?
1. YES 2. NO 3. Cannot answer. YES: What types of improvements?
4.22. The rainwater stored is sufficient for use by the family
1. Only in the rainy season 2. All year round 3. Not enough
4.23. Is the rainwater enough to fill the cistern?
1.YES 2. NO 3. Sometimes

4.24. When the rainwater stored in the cistern runs out, where does the water come from for the family to use?

1.Well 2. River 3. Dam/weir 4. never ends

5. Other: __________

4.25. Does the cistern normally receive water from water tankers during the current drought?

1. YES 2. NO (Skip to question 4.26)

4.26. How many days do you currently receive water from a water tanker?

4.27. Where did this tank car come from?

4.28. What kind of water was he carrying?

4.29. Did you pay or was it free?

4.30. How often do you currently fill your cistern with water tankers?

1. Weekly 2. Fortnightly 3. Monthly 4. Other

4.31. Do you clean the cistern / maintain the cistern?

1. YES 2. NO

4.32 **Do you maintain the cistern?**

1. YES 2. NO

4.33. How often the cistern is cleaned (washed)

1. Half-yearly 2. Yearly 3. don't know

4.34 How is this cleaning done?

1. only washes inside 2. Washes only on the outside

3. Inside and out

4.35. Use a brush to wash

1. YES 2. NO 3. Sometimes

4.36. Do you usually paint the cistern?

1. YES 2. NO 3. Sometimes

4.37. How often the cistern is cleaned (washed)

1. Half-yearly 2. Yearly 3. Don't know

V. HOUSING CONDITIONS

5.1 State of repair of the roof (Interviewer's on-site observation)

1. Regular 2. Irregular 3. Other: __________

5.2 How clean is the roof?

1. clean (skip to question 5.4) 2. Dirty 3. Don't know

5.3 What kind of dirt is found on the roof (Interviewer's on-site observation).
1. Leaves 2. animal faeces 3. Old cans, bottles and tyres 4.Other
5.4 Do you clean and maintain the roof? 1. YES 2. NO
5.5 Do you clean and maintain the gutters? 1. YES 2. NO
5.6 Do you take any precautions to prevent dirt from entering the cistern? 1. YES 2. NO 3. DON'T know
5.7 What method do you use to draw water from the cistern? 1. Bucket 2. Hand pump 3. Other:
5.8 If you use a bucket, is it only used to remove water from the cistern? 1. YES 2. NO
5.9. The system for collecting and storing water in the cistern is inadequate (Interviewer's on-site observation) 1. YES 2. NO (Skip to question 5.12) 3. Don't know (Skip to question 5.12)
5.10. How inadequate (multiple answers) 1. Gutter 2. Pipework (ducts) 3. Crack / Leak 4. Broken lid 5. Bent lid 6. No lid
7. Plaster 8. With various objects inside the cistern
5.11. Gutter system 1. Fixed 2. Mobile
5.12. Animals on the property 1. Cattle 2. goats/sheep 3. pigs 4. Horses 5. Chickens 6. others ________________
5.13. Proximity to animal breeding sites 1. < 10 metres 2. > 10 metres 3. There are no animals
VI. SANITARY SEWERAGE
6.1 Existence of a toilet 1. YES 2. NO (Skip to question 6.7)

6.2 Number of toilets 1. A bathroom 2. Two bathrooms 3. There's a bathroom just for bath
6.3 Location of the bathroom 1. Inside the home 2. Outside the home.
6.4 How often the toilet is cleaned/maintained: 1. Daily 2. Weekly 3. Monthly 4. don't know
6.5 Draining the toilet 1. Dry pit 2. Individual septic tank 3. Septic tank collective septic tank 4. sewage system 5. Body of water (river/stream) 6. dumped in the environment 7. Doesn't know
8.Other:
6.6 Kitchen drainage 1. Dry pit 2. Individual septic tank 3. Septic tank collective septic tank 4. sewage system 5. Body of water (river/stream) 6. Dumped into the environment 7. Doesn't know 8.Others: ________
6.7 Septic tank construction time 1.<2 years 2. 2 years 3. 4 years 4. 6 years 5. 10 years 6. Don't know 7. Doesn't know (Skip to question 6.11)
6.8 Cleaning the septic tank 1. YES 2. NO 3. Sometimes (when it fills up) 4. None
6.9 Conditions of the septic tank 1. Deactivated 2. Inappropriate 3. Adequate
6.10. Disposal of septic tank effluent 1. Soil 2. Body of water 3. Irrigation 4. Others ______________
6.11. Location of the septic tank in relation to the house (Interviewer's on-site observation) 1. forwards 2. Behind 3. Beside

6.12. Location of the septic tank considering the cistern 1. Above 2. Below 3. Beside
6.13. Disposal of septic tank waste (LODO)
1. Backyard 2. Land 3. Rivers 4. Boulders 5. Other:________________
6.14. How the rubbish produced by the family is organised 1. cans 2. Plastic bins 3. Plastic wood 4. Plastic bags 5. No specific container 6. Throws it in the yard
6.15. The family separates rubbish 1. YES 2. NO 3. Only leftover food for animals 4. Just the leaves 5. Other: ________________
6.16. Waste disposal 1. Collected by the town hall 2. Burned 3. Buried. 4. Played in the grounds 5. Thrown into rivers and/or streams 6. Reused.
VII. PUBLIC HEALTH
7.1 In which family members are illnesses most common (Other than diarrhoea) 1. Children < 5 years old 2. Children > 5 years 3. Adults 4. Elderly
7.2 Are there any cases of diarrhoea in the family? How often does it occur? 1. Weekly 2. monthly 3. Half-yearly 4. Yearly 5. None
7.3 In which family members is diarrhoea most common?
1. Children < 5 years old 2. Children > 5 years 3. Adults
7.4 Is there a health worker in the locality? 1.YES 2. NO
7.5 How often do health workers visit? 1. more than once a week 2. Weekly 3. Fortnightly 4. Monthly
7.6 Are there other PSF health professionals (doctors, dentists, nurses...) in the locality? 1.YES 2. NO

7.7 How often do these health professionals (doctors, dentists, nurses) visit the community?

1. more than once a week 2. Weekly 3. Fortnightly

4. Monthly

7.8 The health worker provides clarification on:

1. water treatment 2. Personal hygiene 3. Diseases caused by water

4. Care for the environment 5. Other

7.9 Is there any work in the community on public health?

1.YES 2. NO 3. Don't know

7.10. Type of work

1. Health education 2. Food education

3. Environmental education 4.Others: ____________

7.11. Does anyone in the household take part in the work?

1.YES 2. NO (Skip to Question 7.13)

7.12. After the cistern was installed, were there any improvements in the health of family members?

1. YES 2. NO 3. Cannot answer

7.13. Have any illnesses become less frequent?

1.YES 2. NO (Skip to Question 7.15)

7.14 Do you own a domestic animal?

1.YES 2. NO

7.15. Are domestic animals kept loose?

1.YES 2. NO

7.16. Children have contact with domestic animals

1.YES 2. NO

VIII. HOUSE DIMENSIONS

1. Length of the house (side of the house with guttering):

|__|__|, |__|__| (in metres and centimetres)

2. Width of the house (side of the house that doesn't have a gutter):

|__|__|, |__|__| (in metres and centimetres)

3. Roof height

Height 1 Largest (hl) |_____|_____ |, |______|_____ | (in metres and centimetres)

Height 2 Smaller (h2) | ___| _____ |, | _____ | _____ | (in metres and centimetres) Height 3 If any (h3) |__________| _____ |, | ______ | ____ | (in metres and centimetres)
IX. CISTERN VOLUME
1. Volumetric capacity _________________________ 2. volume observed
X. OBSERVATIONS: 1. Cistern water management 1.1. Inadequacy observed in relation to cistern water management   2. Family Hygiene 2.1. Conditions housing ____ 2.2. Hygiene conditions of the residence ____ 2.3. Hygiene conditions of family members___________

Field Monitoring Form:

Date of collection: ___________________

Observer: ____

Cistern code	Owner

1- Form of withdrawal of water:____

2- N° of people of family:

3- Location of the pit in relation to the cistern:

A. Good (Level below / No pit / Far away)
B. Bad (Level above/next) Other obs.

Proximity to animal breeding sites_______________

4- He received a Kite Car this month: ___________

5- Observed volume ______________________________________

6- State of conservation of the rainwater catchment area

State of repair of the roof ______________________________________

Hygiene (cleanliness) of the roof ________________________________

Type of dirt found in the roof ______________________________________

State of conservation / hygiene / levelling of gutters and conductors

7- - State of conservation of the cisterns:

State of repair of cisterns ___________________________

Inadequate cistern water collection and storage system ___________________

8- - Family hygiene:

Hygiene residence____

Hygiene of family members___________________________

9- - Occurrence of illness in the family

Follow-up questionnaire

Date of collection:	time:
Interviewees:	System/family code

1. Number of people in the family

2. Hasn't it rained in a while?

3. How long has it been since the last rain?

4. Where does water come from in times of drought?

5. When there is no drought, what sources of water does your family (community) have?

6. And today, what kind of water do you have in your cistern?

( ) water cart ( ) rain ( ) dam ( ) rivers ( ) wells ( ) other

7. Do you always fill up with this type of water?

1 Yes 2. No, what other types? Why not?

8. Who is usually responsible for taking the water out of the cistern?

9. PERIOD OF PROLONGED DROUGHT

9.1 Do you have trouble getting water during the dry season?

9.2 During this period, where does the drinking water come from? What does it look like?

9.3 Is this water put into the cistern?

9.4 In times of drought, where does the water come from?

9.5 What kind of water was used to fill the cistern during the dry season?

( ) water cart ( ) rain ( ) dam ( ) rivers ( ) wells ( ) other

9.6 How often do you fill up with water?

9.7 Are there any illnesses in the family during this drought?

9.8 What diseases? (let the interviewee talk about the type, you don't have to give the name of the disease)

9.9 During periods of normal rainfall, before this drought, were there many diseases? Which ones?

9.10 Were the sick people children or adults? How old were they?

10. And the cistern, have you ever washed it?

11. When was the last time you washed the cistern?

12. How did you wash it?

13. Was the cistern full of water during the past rains?

14. When was the cistern full?

15. Was the water stored in the cistern during the rains enough to supply your family?

16. How long did your family use the water from the cistern that accumulated in the cistern during the rains?

17. When the cistern runs out of water, where do you get it from?

During collection: *On-site observation*

1. Water withdrawal method

( ) Bucket ( ) Can ( ) Hand pump ( ) Other

2. Volume of water observed:

3. How much, how many cans per day for consumption and household chores?

4. Amount of water withdrawn per day to make the pots

5. State of conservation of the rainwater catchment area

6. State of repair of the cistern: ( ) good ( ) fair ( ) bad

7. System inadequacies:

( ) cracks ( ) no cover ( ) broken or bent cover ( ) poorly positioned gutters ( ) poorly positioned ducts ( ) pump
manual broken ( ) without manual pump

Other:

8. Family hygiene:

-Hygiene of the home: ( ) good ( ) fair ( ) bad

Other obs:

-Yard hygiene: ( ) good ( ) fair ( ) bad

Other obs:

-Do you have rubbish near the cistern ( ) yes ( ) no

-Do you have animal breeding sites near the cistern ( ) yes ( ) no

-Sewerage within 10m of the cistern ( ) yes ( ) no

-Hygiene of family members ( ) good ( ) fair ( ) bad

OTHER OBSERVATIONS

Appendix

LOICEIRA QUESTIONNAIRE

I. LABOUR DATA COLLECTION
1.1 How long have you been making crockery? ____________________
1.2 Who did you learn from? __________
2.3 Do you have anyone else in your family who works with crockery? 1. YES 2. NO (Skip to question 2.5)
2.4 What are they? 1.Sister 2. Aunt 3. Daughter 4. Mum 5. Sister-in-law 6. Other ____________
2.5 Do you work alone? 1. YES 2. NO (Skip to question 2.7)
2.6 Why do you prefer to work alone?
2.7 Who do you work with? Why do you work with them?
2.8 Do you have anyone interested in continuing the work? 1.YES 2. NO
2.9 Who are they?
III. DESCRIPTION OF THE PLACE WHERE THE CLAY IS REMOVED
3.1. Where is the clay removed?

1 .Rio 2. Dam 3. Dam 4. Lake 5. Other _______________
3.2 Distance from the workplace?
3.3 Who removes the clay? 1. Another person 2. Own person (Skip to question 3.6)
3.4 Is the person paid anything? 1. YES 2. NO (Skip to question 3.6)
3.5 How much did you pay? _______________
3.6 Do you pay any money to the owner of the property? 1.YES 2. NO
3.7 What is the value? _______________
3.8 How long have you been removing clay from the site?
3.9 How many times a week/month is the clay removed from the site? 1.(1) 2.(2) 3.(3) 4.(4) 5.(5)
3.10. Have you ever removed clay from another location?
1. YES 2. NO (Skip to the next questions)
3.9 What location? _______________
3.10. How long did you spend removing the clay from this place? __________
IV. DESCRIPTION OF THE CROCKERY
4.1. Where is the crockery made? 1 Home 2. Shed 3. Co-operative______ 4. Other
4.2 How is the clay prepared? 1. Clay + H2O 2. Clay + Area + H2O 3. Clay + Area + H2O +Colourant
4.3 How many hours/days will the clay be ready? 1. One day 2. Two days 3. Three days 4. Four days
4.4 When do you start working with the crockery?
4.5 How many days a week do you work preparing crockery? 1.(1) 2.(2) 3.(3) 4.(4) 5.(5) 6. (6) 7. Every day
4.6. Is any other material used besides earthenware? 1. YES 2. NO (Skip to question 5.5)
4.7 What material?
1. Sand 2. Dye 3. Other:
4.8. How is the modelling done? 1 Hands 2. Tone 3. Drainage or the use of moulds
4.9. Where does the water used come from? 1. Water tank 2. Cistern 3. River 4. Pond 5. Dam/weir 6. Other
4.10. How many pans are made each week?
4.11. What material is used to fire the crockery?

1. Loving 2. Catingueira 3. Mastic tree 4. Hawthorn
4.12. Why is this plant used?
4.13. Is the wood used bought in? 1. YES 2. NO (Skip to question 5.12)
4.14. How much? ______________
4.15. How many batches are made with (bale, metre, bag)? 1.(1) 2.(2) 3.(3) 4.(4) 5.(5) 6.(6)
4.16. How many batches are made each week/month? 1.One 2. Two 3. Three 4. Four
5. Five 6. Six 7. Other ______________
4.17. How many dishes are burnt in one batch?
V. DESCRIPTION OF THE CROCKERY
5.1 Which parts are manufactured? 1. Stove 2. Cooker 3. Jug 4. Pot 5. Cambuca 6. Moringa 7. Vase 8. Dish 9. Ornamental pieces 10. Other______________
5.2 What are the most produced items? 1. Stove 2. Cooker 3. Jug 4. Pot 5. Cambuca 6. Moringa 7. Vase 8. Dish 9. Ornamental pieces 10. Other______________
5.3 Do you burn your own crockery? 1. YES (Skip to question 5.5) 2. NO
5.4 Who burns your dishes?
5.5 What happens to the crockery once it's finished? 1. Straight to the sale 2. Co-operative 3. Other Ceramist 4. Other ______________ Note: Answers 2, 3 and 4 skip to question 5.7)
5.6. Where are the crockery sold? 1. Trade Fair 2. Craft fair 3. Shop
5.7 Which days are the crockery sold?
5.8 What is the value of the parts? 1. Stove ________ 2. pan________ 3. jug________ 4. Pot_____ 5. gourd ________ 6. moringa ________ 7. Vase____ 8. plate ________ 9. ornamental pieces ________ 10. Other _____
5.9 Have you made any improvements to your family as a result of working and selling your crockery?

INFORMED CONSENT FORM

Dear participant:

You are being invited as a volunteer to take part in the research:

Your participation involves (mention the procedure; e.g. an interview, which will be recorded if you allow it).

Participation in this study is voluntary and if you decide not to take part or wish to withdraw at any time, you are free to do so.

When the results of this research are published, your identity will be kept strictly confidential. All information that could identify you will be omitted.

Name and signature of supervising teacher/supervisor

I consent to take part in this study.

Participant's name and signature   Place and date

Printed by Books on Demand GmbH, Norderstedt / Germany